WHAT EVERY POTENTIAL HOMEOWNER SHOULD KNOW ABOUT CONSTRUCTION

Residential construction information and details that every potential homeowner should be familiar with.

by DOUGLAS HEDLUND
Vol. 1 / General Construction

Published by:
CONDATA COMPANY / TUCSON, ARIZONA

"The Authors Appreciation to . . .
Mr. C. Foster for his thorough technical review and comments."

What Every Potential Homeowner Should Know About Construction.

Published by
Condata Company • Publishers
P.O. Box 32137
Tucson, Arizona 85751

All rights reserved. No part of this book may be reproduced or transmitted in any form or by any means, electronic or mechanical, including photocopying, recording or by any information storage and retrieval system without written permission from the author, except for the inclusion of brief quotations in a review.

Copyright © 1989 Douglas E. Hedlund
First Printing
Printed in the United States of America

ISBN 0-9621420-0-X

CONTENTS PAGE

PREFACE, WHAT THIS BOOK IS *NOT* ABOUT! 1
INTRODUCTION 3
CHAPTER I   REGIONAL DIFFERENCES IN
              RESIDENTIAL CONSTRUCTION 7
CHAPTER II   THE ROLE OF BUILDING CODES AND
              BUILDING OFFICIALS 13
CHAPTER III   SITE PREPARATION 15
CHAPTER IV   FOOTINGS AND FOUNDATIONS 25
CHAPTER V   FLOOR SYSTEMS 41
CHAPTER VI   EXTERIOR WALLS 61
CHAPTER VII   WINDOWS, GLASS, DOORS AND
              HARDWARE 73
CHAPTER VIII THE ROOF STRUCTURE 87
CHAPTER IX   ROOFING AND FLASHINGS 95
CHAPTER X   INTERIOR CONSTRUCTION AND
              FINISHES 115
CHAPTER XI   CABINET WORK AND OTHER BUILT-INS 139
CHAPTER XII   CHIMNEYS AND FIREPLACES 141
APPENDIX 'A' STATE LISTING OF MODEL
              BUILDING CODES IN USE 149
APPENDIX 'B' SEPTIC SYSTEM DATA FOR PIMA
              COUNTY, ARIZONA 150
APPENDIX 'C' MINIMUM RECOMMENDED 'R'
              VALUES FOR INSULATION 151
GLOSSARY 152

PREFACE ... WHAT THIS BOOK IS *NOT* ABOUT!

It is not the purpose of this book to advise nor guide persons thru the complex and involved process of determining what type, style or size house should be built, by whom, at what cost, nor under what terms and conditions. Nor does this book provide advice for persons wishing to undertake *do-it-yourself* construction of their home.

Rather, this book is intended for persons who are unfamiliar with construction methods and details, but who will at some time be contracting for, or purchasing, a home to be built for them by other qualified parties (Builders, Developers, Contractors, etc.). For them, this book proposes to explain many typical conditions and details which will likely be encountered; thus providing an awareness and capability to evaluate progress and results.

This book proposes to provide accurate and authoritative information regarding the subject matter. However, it is sold with the absolute understanding that the author and publisher are not thereby rendering legal, architectural or other professional services.

The author and the publisher accept neither liability nor responsibility to any person or entity with respect to any loss or damage caused, or alleged to be caused, either directly or indirectly, as a result of the information contained herein.

INTRODUCTION

Hundreds of thousands of new homes are constructed each year in this country by builders and developers responding to the housing needs of a growing, maturing population. The vast majority of persons who purchase those homes are not knowledgeable in the technicalities of building construction. They simply do not have expertise in construction principles, the proper application of materials and components, in field techniques, what constitutes good workmanship, nor the role of local codes and conditions. This is certainly understandable, since this is a very specialized and technical field. However, all of these factors significantly impact and influence the success of the building project. As a result, the average soon-to-be homeowner is not equipped to recognize, nor identify, potential problem-laden issues which can take place daily during the construction period; issues and problems which can have substantial impacts on the quality, durability, and cost of future maintenance of the finished building,—not least of all, on their personal satisfaction with it.

This certainly represents an unfortunate situation for persons who are making such a major investment and long-term commitment.

Being convinced that some comprehensive information on the subject is therefore needed and could make a world of difference to this audience, it is hereby the Author's intent to:

1) Provide basic information about the materials, components and systems which are typically assembled under field conditions into residential building construction;
2) Describe principles and suggest details which can result in quality construction;
3) By editorial use of the device: (**CAUTION**), to focus attention throughout the text on problem areas, pitfalls, and other potential deficiencies which can, and do, occur;
4) Provide the reader with an improved basis to recognize problems, then make judgements about impacts and alternatives; so that if necessary, action can be taken to achieve corrective measures before deficiencies and less-than-satisfactory conditions are buried or covered up by subsequent work.
5) Overall, to assist the reader in becoming more able to deal

with the construction process, **on an informed basis**.

Extensive use is made of graphic diagrams and pictures throughout, to illustrate the issues or details being discussed. The illustrations *are not* intended to be sufficiently finished nor complete to be usable for actual construction purposes. Instead, each is directed at pointing out, by illustration, a particular construction principal, or a *pitfall*. Most are drawn in the same format used in actual building plans and documents, but are kept simplified to best illustrate the particular issue being discussed, and to not discourage the reader by being unnecessarily technical.

It is definitely not intended that the use of this book in any way minimize, substitute for, nor negate the need for Architects and other professionals. Their expertise and services should be sought by those persons who wish to have special, customized and unique design solutions provided to meet their special needs. On the contrary, the author is keenly interested in **encouraging** the public to seek out professional help and guidance in the building design and construction process, for he sincerely believes that the most interesting, unique and satisfactory building solutions result from the use of trained professionals.

The author, however, recognizes the reality that a substantial number of new private residences are developed and constructed directly by builders or developers without the assistance or services, to the purchaser, of an Architect. It is to those persons that this book is intended, in the expectation that thru the use of this information, those persons will be much better equipped to recognize what's taking place during the construction process; and, thus have a better basis to deal with problems, rather than be subject to consequences later on.

This book, therefore, is directed to those persons who will eventually be securing a new conventional residence which is constructed by some other party, as well as to those who contemplate having additions or alterations made to existing conventional residences by contractors or builders. The construction principles and issues apply to either scenerio.

This book deals with *general construction* aspects of residential building; in other words, physical *bricks and mortar* items such as foundations, walls, floors, roofs, and finishes. Typical Mechanical, Plumbing and Electrical systems are not included herein. A subsequent text is being considered by the author to cover these subjects. This is done with the conviction that many people would prefer to restrict their level of study to only the more obvious physical features of their residence and not be burdened with more detail on those systems which are more technical in nature, and physically less obvious. This also allows the reader to make a choice as to the level of interest and involvement in the total building picture, by dividing a very comprehensive subject into two volumes.

CHAPTER I ... REGIONAL DIFFERENCES IN RESIDENTIAL CONSTRUCTION

There are many differences—or variations—which occur in construction as a direct result of the region, or part of the country, in which the building is located.

Many of these differences evolve directly and naturally from the particular conditions of climate, moisture, kinds of materials available, etc. which exist in the particular region, and can therefore be considered truly indigenous, functional differences. A good example of this is the use of sun-dried mud adobe in the southwestern United States. In this region there is readily available the particular adobe clay soil which was long ago discovered to be easily made into usable and inexpensive building blocks. This region also experiences significantly less rainfall, which would otherwise contribute to an accelerated deterioration of the relatively unstable adobe mud blocks. Another significant difference can be pointed out in those areas of the country where frost and freezing conditions penetrate into the soil, necessitating that footings and foundations be set into the ground to a depth below that which frost penetrates, so as to avoid the upheaval and movement which can take place from freeze-thaw action. This can require footing depths of 3 to 4 feet, and more, below ground surfaces. See (Figure 1).

A regional by-product of the need for deeper foundations is the tendency for residences in those colder areas to have below-grade basements, or cellars, because they are logical extensions of the deeper excavation process. Added space is thereby gained, at not-too-significant added cost. See (Figure 2).

Basements are often not without problems, however, because those regions usually have higher ground water levels which result in a tendency for basements to have water leaks and higher humidity conditions. Many methods and details for treatment of the water/moisture problem have resulted, including application of waterproofing to interior and exterior surfaces of basement walls, under-floor drainage systems, underfloor waterproofing membranes, sump pump systems, etc. See CHAPTER IV for more discussion on this problem.

In contrast to the deep foundations of cold regions, the warmer southern and southwestern areas of the country usually are

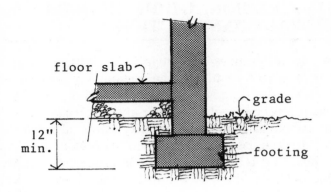

FOOTINGS IN WARM CLIMATES

FIGURE 3

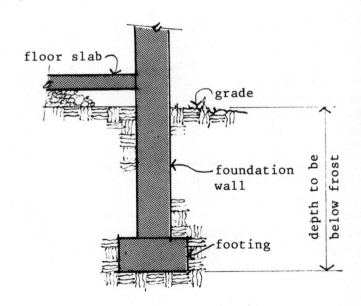

FOOTINGS IN COLD CLIMATES

FIGURE 1

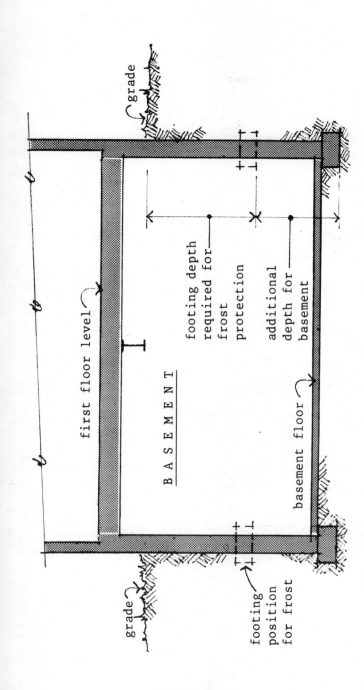

SECTION THRU A BASEMENT, COMPARING COLD CLIMATE FOOTING DEPTHS

FIGURE 2

devoid of frost penetration into the soil, and so footing depths can be quite shallow—a common minimum depth being as little as one foot below natural ground level to bottom of the footing (Figure 3). Therefore, in these areas there is much less incentive to create deep excavations for below-grade basements, and so they are infrequent. Further, the main floor is usually a concrete slab placed directly at grade on prepared soil bases, since there is no excavated area below, and therefore no need to form a structural system capable of spanning open excavated areas. Contrast this with the usual joist and beam system commonly used over excavated basement spaces (Figure 2).

In colder regions, on occasions when basementless slab-on-grade construction is used, it is not common practice to pour the floor slab integrally with the foundation wall, as is often done in the southern regions (Figure 4).

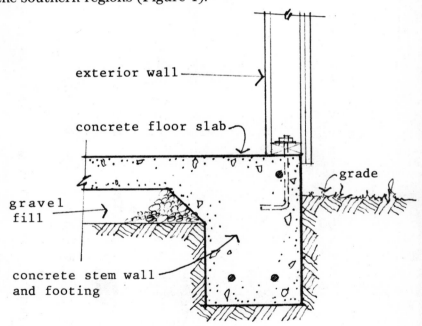

INTEGRAL FLOOR, STEM WALL AND FOOTING
IN WARM CLIMATES

FIGURE 4

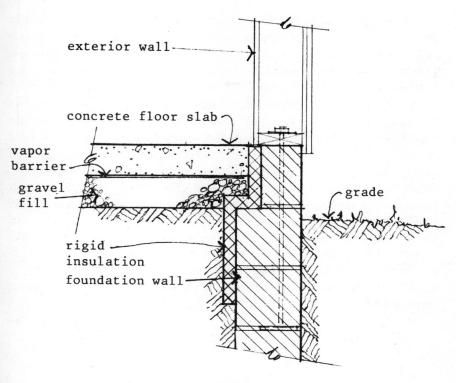

FLOOR SLAB INSULATED AT EDGES
IN COLD CLIMATES

FIGURE 5

Rather, it is desirable—in fact necessary—to isolate the floor slab from the exterior wall via an insulated separation, in order to minimize heat loss from the warm interior to the cold exterior ground thru the perimeter edge of the slab and foundation (Figure 5).

Another indigenous difference is the widespread use of stucco as an exterior finish in the south and southwest; a material only infrequently found in the colder regions of the country. The reason for this, is that stucco is highly susceptable to damage from penetration of moisture due to frequent wetting, then being subject to freeze-thaw cycles which deteriorate it, cause it to loosen from its base, crack and spall (fall off).

Chapter 1: Regional Differences in Residential Construction

Many regional differences in building features exist simply because of design or appearance characteristics which have developed and become widely accepted in particular regions. These features contribute to the special uniqueness and character of that region and often provide a desirable identity or traditional quality. Examples of these are: the *Cape Cod* designs found frequently in the northeast; the *Santa Fe* style of the southwest; the *Raised Ranch* of the east and mid-west; the use of clay tile roofing in the south and southwest; the design features attributed to the *California Ranch* style of the west coast, etc. Some of these design or appearance characteristics result from functional requirements as well. An example is the widespread use of steeper-pitched gable roofs in the cold and rainy areas of the country, in order to more quickly shed the frequent rains, and to minimize large build-ups of snow and ice. Contrast these with the frequent use of so-called *flat* roofs of the Santa Fe style or other southwestern *territorial* style houses, where those climatic requirements are greatly diminished.

CHAPTER II . . . THE ROLE OF BUILDING CODES AND BUILDING OFFICIALS

Building codes are documents which, by virtue of their adoption by a town, city, county or state, become law; and in which, are set forth the conditions, details and/or results that are required in the construction of buildings. Their purpose is to provide standards and requirements of construction which are the **minimum** allowable in order to safeguard the health, welfare and safety of building occupants and the general public. Protection of property values is secondary.

Every major metropolitan area as well as most minor towns, cities and counties of the country have adopted some sort of building code. However, not every location in the country is governed by a code. There are several model codes which have been written by various expert groups or organizations, and are made available for adoption and use by governmental entities. An example of such a model code, is the UNIFORM BUILDING CODE written by the International Conference of Building Officials, and currently in effect in most of the western states of the country. There are others. For a listing of some of the more prevalent codes in use see **APPENDIX 'A'**.

Building codes are of two general or generic types; i.e., *Specification* or *Performance*. A Specification code is one which provides specific requirements as to the allowable methods, systems or details which are to be used, the types of materials, their allowable strengths, etc. The UNIFORM BUILDING CODE is an example of a Specification Code. Conversely, a Performance code describes conditions or results which must be met and the generally acceptable and recognized standards that can be used as guides to achieve and test those results.

All building construction in the area of jurisdiction must conform to the requirements of an applicable Building Code. Every entity which has a Code will have a department, agency or official responsible for enforcement of its provisions, commonly called the Building Official, Building Inspector or similar title. These officials have the responsibility of reviewing plans and specifications for proposed construction to determine conformance of the documents with the adopted Code, issuing building permits to allow construction to proceed, and performing certain field inspections as they deem necessary to assure themselves and the government body under whom they work that the construction is proceeding in conformance with the Code. Upon satisfactory completion of the construction, they will usually issue a Certificate of Occupancy, which allows utility systems to be turned on and building owners and occupants to legally use and occupy the building.

Building inspectors make periodic visits to the on-going construction, usually at very specific stages of project progress development. Examples of typical stages are: At the completion of site preparations for footings and foundations, but just prior to the placing of concrete; At the completion of rough framing, prior to any surfacing or enclosure; At the completion of rough-in of electrical wiring, etc.

(**CAUTION**) The intent of the inspector's visits is not necessarily to ensure the contractor's conformance with all specification requirements nor to generally protect the owner's interests; but rather, to determine conformance to the Code. Many things can be done, or not be done, which will either escape the attention of the official inspector, or for which he/she has no interest or jurisdiction. The uninformed owner or buyer is then left entirely to the practices, intentions, knowledge and good faith (or lack thereof) of the builder plus his sub-contractors to produce the expected (and paid for) results. That is why it is in your best interests as the Owner and Occupant to become informed, to thus be more able to observe, understand and judge what's going on in the complicated construction process.

CHAPTER III . . . SITE PREPARATION

The design and engineering of overall site features of a residential development or sub-division is the responsibility of qualified Architects, Civil Engineers and other professionals. They must determine the layout of lots, the overall surface drainage patterns, roadway locations and details, utility systems (water, gas, sewer, electricity, etc.), the general locations and positioning of buildings to assure conformance with zoning requirements of density, setbacks, heights, etc., and other similar project-wide considerations. It is therefore beyond the scope and intent of this book to discuss those development-level planning factors. Instead, our comments regarding the site will focus on the immediate lot or site for an individual residence project, to provide some guidelines and factors to consider.

DRAINAGE:

If adjacent abutting property in its natural state is higher but pitched towards yours, some modifications such as a drainage swale (ditch) or other grading re-configuration on your property may be necessary to divert surface water so it will not be a hazard or detrimental factor to the use of your land. Conversely, any alterations to the topography of your site must not thereby divert water flow to your neighbor. If the topography of either property must be deliberately altered (re-graded) because of the development or building process, the resulting configuration must not divert surface flow or drainage to the other property if that flow did not exist in the natural state.

It is important to be sure that placement of buildings and grading of land adjacent to them be such that surface water flow is directed away from—instead of towards—the buildings. Elevations of any floor slabs at grade should, at the **absolute minimum**, be 6 inches above the highest adjacent grade. Any slab at grade which ends up with its top elevation at or below the adjacent ground has a very high potential for receiving standing or flowing water during storm conditions.

Sloping or hillside land may require special treatment in the form of cuts, fills and/or terracing to divert and control run-off water. On the sides of the building which face rising slopes, a drainage swale (ditch) should be created at the base of the main slope, and fill material then be placed so that there is a rise of grade from the swale back up to the building, thus forcing surface drainage away from the building, towards the swale. See (Figure 6).

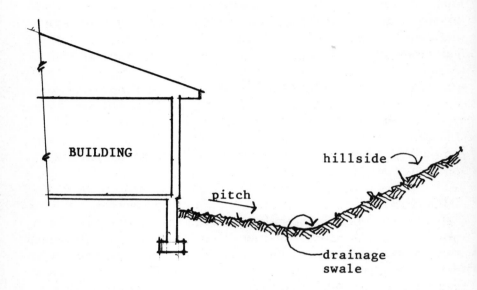

DRAINAGE SWALE ALONG HILLSIDE

FIGURE 6

GRADING, FILL AND BACKFILL:

It is highly desirable to minimize changes in the topography of a building site. Whenever possible, the building should be built on natural grades rather than modified conditions. This is because all modifications alter the load carrying capacities and drainage characteristics of the soil.

In the area to be occupied by the building—including all outside walls, paved areas, driveways, etc.—all plant stumps and roots should be removed to a depth of at least 12 inches below ground surface.

(**CAUTION**) Be especially concerned about areas within the actual profile of the building which are proposed to receive fill or disturbed soil. Once disturbed or excavated, the compaction and density of soil is dramatically reduced, because when loosened the void-to-solid ratio of the soil is increased. Filled or backfilled areas must therefore be placed and compacted by special means, with special care. Fill/backfill material should **not** contain clay, topsoil, loam or construction debris. It is especially important that scraps of lumber, form boards, grade stakes or wood debris of any kind not be placed into, or left in, fill and backfill areas. This is to prevent voids left by rot and decayed material, and especially to prevent attraction of subterranean termites.

Backfill material should be granular in texture and reasonably permeable to water. Fill/backfill must be placed in layers (lifts). Depending on the type of soil, each lift should generally not exceed 6" to 12" in thickness, or depth, and each lift then be compacted by special mechanical equipment to achieve at least 90% to 95% of its original density when dry. Unless this procedure is followed, filled or backfilled areas will eventually settle, often severely and usually uneven in amount at different locations, with very unfortunate impacts on the building. These can include broken/failed footings and foundation walls, cracked floors, ruptured pipes, warping, distortion and cracking of walls and ceilings, etc. When fill is placed in thick lifts, any efforts to compact the material will produce results only in the upper portions of the material, not throughout the entire depth of the lift. It is simply not possible to achieve true compaction to the proper density when placed in a single lift. Unfortunately, this is a too common practice, especially in the backfilling of trenches or against foundation walls. The consequences of subsequent settlement are obvious.

The contractor or builder should be required to have all areas of fill tested for proper compaction by an independant testing laboratory, starting at the bottom of the excavation. You have every right to insist on seeing the results of those compaction tests; and, if you have any doubts or concerns about filled conditions in your building site, insist on testing and on a professional's interpretation of the results.

TOPSOIL:

Topsoil is the top few inches (usually not more than 6 to 8 inches) of the ground surface, rich in organic materials, having good drainage characteristics. If the building site has topsoil it should be saved for future use. Topsoil definitely should not be used for fill, backfill, nor be removed from the site. Have the builder *stockpile* it out of the way of construction, for future re-use for lawn areas, plantings, gardens, etc. He simply bulldozes it off into a storage pile. Nature has taken millions of years to create the unique materials composing topsoil, and it should not be ignored nor wasted. However, not all sites have topsoil of the quality and characteristics we are referring to. If you are personally unable to make an identification, have a Landscape Architect or specialist look at it and advise you.

CLAY:

Clay can be the most troublesome material encountered on a site. Clay is impermeable to water and so makes drainage and percolation of water into the ground difficult; it is usually compressable and so allows more or less settlement than other soils. It also can be *expansive*, i.e., subject to an increase in volume when saturated with water. Clay should not be used for fill or backfill because of these characteristics. Generally, building footings and foundations should not be placed directly on clay. However, clay can be useful as a means to control and limit water percolation into the ground, by its placement as the top layer over backfill. This can be effective along basement walls to keep water from collecting in the former excavation *pocket*, and thus become a problem for the basement spaces.

RETAIN AND PROTECT PLANTS:

Try to retain and maintain existing trees and other major plants, even if it requires slight adjustments in building placement to do so. These trees and plants are almost always an asset to the surroundings; useless destruction is always unfortunate. Plants are expensive, and take substantial time to re-grow to maturity. A good landscape plan will utilize existing major plants rather than destroy them.

Be alert to the builders' efforts to physically protect trees and their root systems from construction activity. If they are close to construction areas they may require wrapping or other physical buffers to keep machinery and equipment from physically damaging them. Roots, if exposed, should be re-covered as soon as possible before dry-out occurs, and then be kept moist.

SEPTIC SYSTEMS:

If public sewers are not available, you will likely have an on-site septic system for sewage disposal. The physical location and relationships of a septic system to the buildings, to other site improvements, and especially to any water wells on the site or on adjacent property, requires special consideration, and has legal constraints. Local laws will usually be in force which dictate minimum distances from various features, especially water wells. This is to minimize the potential for contamination of ground water and wells, due to seepage of the effluent (waste water) from disposal field or pits. See **APPENDIX 'B'** for an example of setback/distance requirements for Pima County, Arizona.

Be sure that the sizing of the septic system tank and disposal field has been based on local regulations, and on *percolation tests* done at the site. These are simply holes dug to a certain size and depth in the earth, filled with water several times, and the time for the water to drain down recorded. This information is then used as a basis for calculation of the required surface area of the disposal system. This process measures the permeability, or resistance, of the soil to water absorption/percolation, a key factor to the functional life of a septic system.

A grave misconception about septic systems is that they are permanent systems that, once installed and operating, never need to be maintained, checked on or cleaned out. This is **not true**. The septic tank is a receiving and holding vessel in which certain bacterial action takes place on the contents, and from which fluids drain out to the disposal system for percolation into the soil. However, the bacterial action in the tank does not completely decompose all material that enters it, converting it all into liquid which the disposal field can leach away. Instead, a *scum* or foamy mixture of gasses, bacteria, and suspended solids forms and floats on top of the tank contents; heavier solids, called *sludge* settle to the bottom of the tank. The scum and sludge gradually accumulate and fill the tank. If these are not periodically removed from the tank they eventually flow into the outgoing (effluent) and incoming (supply) piping, clogging and ruining the system. These materials must be removed by pumping out the tank every 3 to 5 years, maximum, depending on the number of building occupants. This is why septic tanks have large removable access covers, and why a tank location must be documented, be accessable and not be too deep in the ground. For a diagram of a typical septic system, see (Figure 7).

TREATMENT FOR SUBTERRANEAN TERMITES:

Subterranean or ground-nesting termites, are the most destructive insect pests of wood. They are found in all States except Alaska, but are most common and aggressive, and hence most destructive, in the more temperate parts of the country. See (Figure 8) for an indication of the relative density of termite activity throughout the United States. Subterranean termites thrive in moist, warm soil containing an abundant supply of food in the form of wood or other cellulose material. They often find such conditions beneath buildings where the space below the first floor is poorly ventilated and where scraps of lumber, sawdust, form boards, stumps or roots are left in—or on—the soil. Most termite infestations in a building occur because wood used in its construction touches, or is close to, the ground. This could be at sills, porches, steps, wood posts, terraces or decks. In addition, cracks or voids in foundations and concrete floors make it easy for termites to reach wood that does not actually touch the soil. Any crack 1/32" or greater will permit the passage of termites. Termite activity can exist even in colder northern areas when soil within or adjacent to heated basements is kept warm throughout most of the year.

There are a series of construction practices and details which can seriously impact termite access and damage to residential buildings, as follows:
1) Do not allow wood products to be left, or buried, in the soil under or near the building.
2) Keep drainage away from the building and from collecting near or under it.
3) Make footings and foundation (or basement) walls as impervious as possible to termites by striving to eliminate voids, cavities and cracks thru which the insects can penetrate and use as paths for access to wood members.

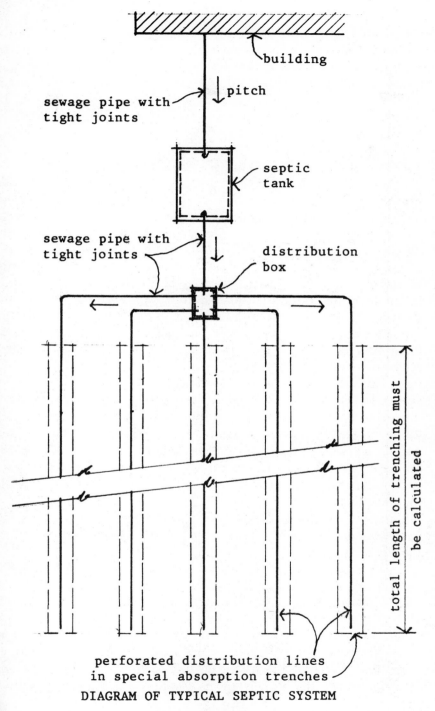

DIAGRAM OF TYPICAL SEPTIC SYSTEM

FIGURE 7

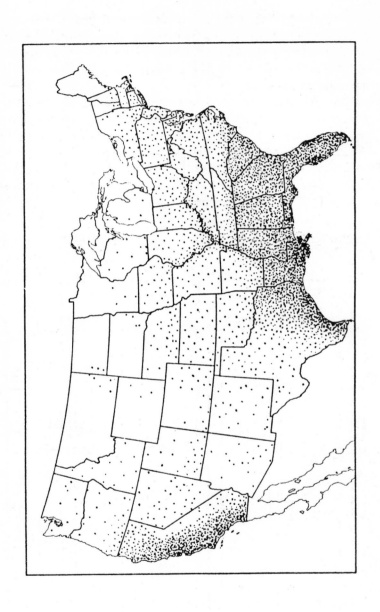

DENSITY OF STIPPLING INDICATES RELATIVE HAZARD
OF TERMITE INFESTATION IN THE UNITED STATES

FIGURE 8

Reinforced poured concrete footings and foundation walls offer the best chances of achieving this. Walls of hollow masonry block should have solid poured concrete caps and/or have the hollow cores of the top course completely filled with cement grout.

4) Consider the use of metal termite shields at the top of foundation walls, just below any wood construction. See the discussion and details on this in CHAPTER V..FLOOR SYSTEMS.
5) Maintain certain minimum distances between finished grades, or soil levels in crawl spaces, to all wood construction above. See (Figures 9 and 10).

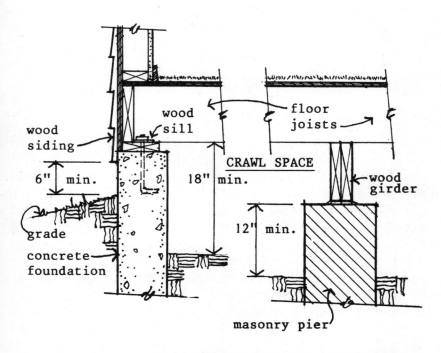

CLEARANCES OF WOOD FLOORS TO SOIL, FOR TERMITE INSPECTION AND CONTROL

FIGURE 9

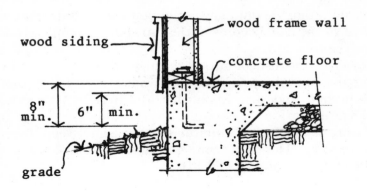

**WOOD CONSTRUCTION ON CONCRETE SLAB,
FOR TERMITE INSPECTION AND CONTROL**

FIGURE 10

6) Chemically treat the soil around and under foundations and grade-supported floor slabs of buildings.

The chemical treatment approach is mandated in certain western and southern states. Certain toxicant chemicals are approved and in use for this purpose. Specific solutions and concentrations of these chemicals must be applied to the exposed soil under concrete floor slabs, along and under the foundations of wood framed buildings.

Specific details and recommendations regarding these preventative measures are well described in a publication of the United States Department of Agriculture, Home and Garden Bulletin Number 64, entitled : SUBTERRANEAN TERMITES, THEIR PREVENTION AND CONTROL IN BUILDINGS. This document is the authority referenced for chemical treatment requirements by certain governmental jurisdictions in the state of Arizona which use the UNIFORM BUILDING CODE.

CHAPTER IV... FOOTINGS AND FOUNDATIONS

It's an old perhaps trite, but true saying that: *"A house is only as good as its foundation."*

A foundation consists normally of two components: the footing(s) and the foundation wall(s). Footings are generally of poured concrete. The foundation walls are usually of poured concrete or concrete block; stone is found occasionally in the walls of older houses; brick is also occasionally used. There are even foundations built of specially treated wood, but these are more experimental and rare, and as such will not be further discussed herein.

FOOTINGS:

Footings are the structural elements which receive loads from all of the other portions of the building (walls, floors, partitions, roofs, etc.), and transmit or spread those loads uniformly to the soil. Different types of soils have different load carrying capabilities, technically called *load bearing capacity.* Solid bed rock has the highest load bearing capacity, varying from a low of 2 tons per square foot, increasing to very significant figures depending on the type of rock. By contrast, clay or silty clay may have a load bearing capacity as low as 1/2 ton, or less, per square foot. For major structures with significant loads, such as multi-story buildings, an extensive sub-surface investigation and testing procedure is required to investigate, analyze and determine the soil characteristics and its load bearing capacities. These extensive investigations are not usually undertaken for very light buildings such as the average residence; however, some investigation must be conducted to categorize the type(s) of soil to be encountered, and to ascertain if any serious problem areas exist, such as high water tables, very poor quality soils, extreme differences in kinds of soil, presence of fill, rubbish or land-fill/dump areas (yes, it does occur). A common method of exploration for residential construction is to dig open pits at least to the depth of the deepest expected footing levels and have a soil expert visually examine the soils encountered. Oftentimes, samples will be taken for laboratory analysis to further assist in the determination of soil suitability. In any event, the builder **must** know by some exploratory basis, what he is expecting to encounter for sub-surface conditions. His responses to some simple inquiries on your part should confirm his basis, and dispel any doubts. Building codes will assign specific (and usually very conservative) load bearing capacities to be used for various soils if testing is not undertaken.

(**CAUTION**) Footings must be deep enough into the earth to be below the levels of any possible frost penetration or frost action. As explained in CHAPTER I, footings which are subject to frost action are subject to movement and differential settlement.For examples of footings and foundation details in regions which are subject to ground freezing, see (Figures 1 and 2). For examples of those items in the warm south and southwestern areas of the country, see (Figures 4 and 11).

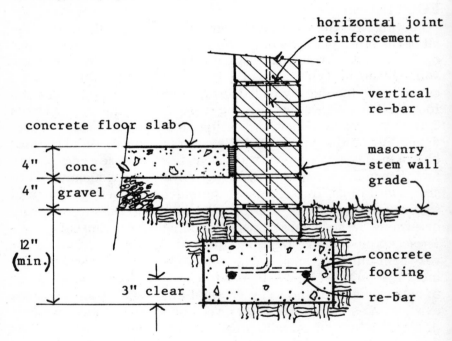

A TYPICAL FOOTING AND FOUNDATION DETAIL
IN WARM CLIMATES

FIGURE 11

Footings are generally of two basic types: *Continuous wall* footings which are long continuous lengths having one constant cross-sectional profile (Figures 11 and 12);

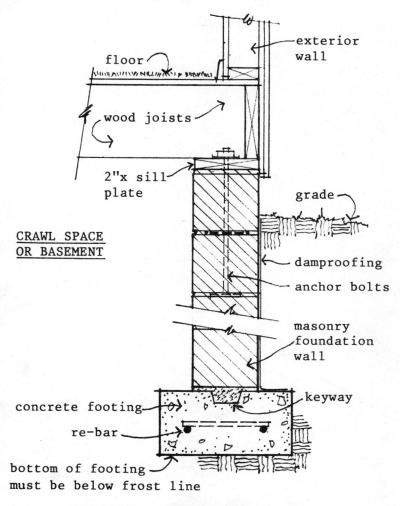

A TYPICAL FOOTING AND FOUNDATION DETAIL
IN COLD CLIMATES
FIGURE 12

or *individual* column, post or pier footings which are usually square or rectangular in plan, by several inches in thickness (Figure 13).

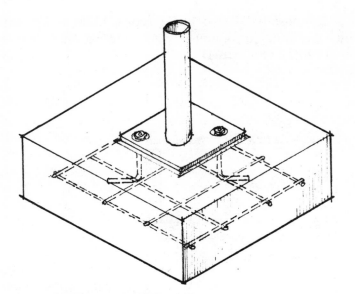

COLUMN FOOTING

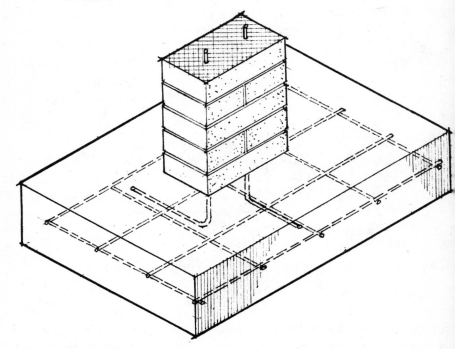

PIER FOOTING

FIGURE 13

Continuous wall footings in residential structures are usually a minimum of 12 inches to 16 inches in width, but always at least 6 to 8 inches wider than the wall resting on it, and a minimum of 8 inches thick. If the foundation wall is thicker than 8 inches, the footing should be also. The actual width and thickness are subject to determination by calculation based on the imposed loads vs. the determined soil bearing capacity. However, except for unusual soil conditions, the above rule-of-thumb sizes will usually suffice for normal single story residences.

The top surface of continuous wall footings should have a depressed slot, called a *keyway*, cast into them, for the purpose of providing improved connection between the foundation wall and the footing. This adds resistance to movement of the foundation wall during backfill operations, as well as from lateral or sideways forces which take place from the earth itself. See (Figure 14)

Column or pier footings' sizes should at all times be determined by calculations. Their plan dimensions, thickness and reinforcing requirements cannot be rule-of-thumb applications because the soil bearing loads ultimately assigned must be similar to those assigned to all other footings, to avoid differential settlement of the building. Individual footings are subject to bending forces in all directions, which must be calculated to determine necessary thickness and the amount of steel re-bar required.

REINFORCEMENT:

(**CAUTION**) Footings should always be reinforced with deformed steel bars (commonly referred to as *re-bar*), made especially for reinforcement of concrete. In continuous wall footings, there should be a minimum of two #4 bars running continuously along the length of the footing longitudinally, plus #4 bars at some selected spacing across the longitudinal bars, to properly space and position the long bars. Bar numbers refer to the number of 1/8th's of an inch in their diameter; therefore, a #4 bar is 4/8th's or 1/2 inch in diameter. Re-bars serve to strengthen the concrete by giving it ability to resist tension—or pulling—forces, and thus create beam action so that footings will span soft spots or voids in the soil without breaking or causing differential settlement. In column or pier type footings, bars should be

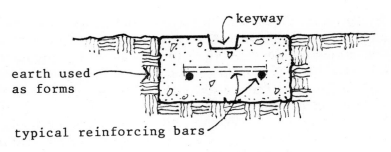

TRENCH FORMED FOOTING

FIGURE 14-B

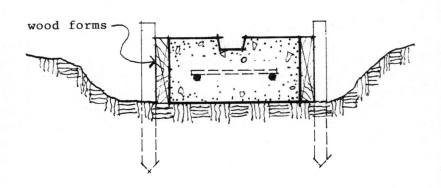

WOOD PLANK FORMED FOOTING

FIGURE 14-A

required at certain spacings across both the footing width and length, based on calculations made for bending stresses placed in the footing by the column or pier loads, resisted by the resulting forces being pushed back by the soil. The bars in footings must be placed so that no face of any bar is closer than 3 inches to faces of dirt at the bottom or sides of the footing. In order to hold the bars in position, special metal support or *chairs* are installed at specific spacings. These rest on the soil and support the main bars. An alternative method of bar support often used by builders, is to drive short sections of re-bar into the soil at the bottom of the footing, attaching the main bars to them. Special plastic caps, or separators, must be used to prevent metal-to-metal contact. This prevents water from migrating up the support chairs, or bars, and rusting out the main bars.

(**CAUTION**) Pieces of masonry, brick, concrete block, stone or broken concrete **must not** be used to support or position re-bars. This is because all masonry materials will absorb water from the underlying soil, which will in turn be drawn up to the re-bar, rusting it out and destroying its value.

(**CAUTION**) All bars should be new, clean and free of oil, grease, rust, scale, dirt or any other material which would interfere with their ability to bond to the concrete.

FOOTING FORMS:

Wherever possible, footings should be cast into side forms of wood or steel surfaces. This assures accurate and consistent dimensional control of the elevation, width and thickness of the footing. The forms are later removed. See (Figure 14-A). Occasionally the soil is stable enough to be suitable for using the profile of the dug trench as the form (Figure 14-B). This is less desirable however, because of the less uniform dimensional control, the rough and uneven nature of the earth faces, and the tendency for earth to loosen and fall into the concrete during the pouring process.

(**CAUTION**) Prior to pouring concrete, it is essential that the soil at the bottom of the footing trench be level, undisturbed virgin soil, or soil which has been properly compacted to high acceptable densities, and be completely free of loose soil, debris, and water. Any violation of this requirement will impact directly on the integrity and quality of the footing.

VERTICAL RE-BAR:

Some regions of the country, due to local code requirements, require masonry foundation walls to be reinforced both horizontally and vertically with steel reinforcement. In order to tie these walls to the footings, vertical re-bars must be placed in the footings at intervals, and of sizes, to match those required in the walls above. These bars should be bent at their base to provide a foot—or hook—for proper imbedment and bond into the footing. These bars should be wire tied to the main footing bars, and be temporarily aligned and held in proper vertical and horizontal position until the concrete is set.

STEPPED FOOTINGS:

In order to maintain the required minimum depth of footings below ground surfaces which slope, footings are stepped down in elevation, not unlike stair steps. The maximum incline or ratio of rise-to-run of such steps is 1 to 2. Any less ratio or incline is O.K., for example, 1 to 3, 1 to 4, etc. See (Figure 15). Footing bottoms should **never** be sloped to accommodate changes in depth.

CONCRETE:

Concrete is a mixture of stone, sand, portland cement and water. The proportions of each material are critical to the strength and *workability* of the concrete. The mix hardens by absorption of the water into the portland cement to hydrate it, as the mix *sets up*. Development of full working strength takes a minimum of 28 days. The most critical factor which affects strength is the ratio of water to cement. Higher cement content for the same (or less) water, increases strength.

(**CAUTION**) For this reason, additional water should not be added to already mixed concrete. If for some reason the mix is just too *stiff* (or dry) precluding its proper flow into the forms, or is necessary to slightly improve workability or plasticity, only an **absolute minimum** quantity of water should be added. Small variations in water content make large differences in workability, but also in strength. Concrete should never be so wet so as to flow like soup or water. It must, however, be able to fill all voids of the forms without the ingredients being segregated. A field measure of the stiffness, or water content, of concrete is the test for *slump*. This tests how much a cone of fresh concrete will sag (slump) under its own weight. The maximum allowable slump should be from 4" to 5".

(**CAUTION**) Concrete must not be dropped from great height. 3 to 4 feet of drop is the maximum. For placement from greater heights, troughs, chutes, tremie or mechanical pumping must be used. Dropping causes separation of the stone aggregate and

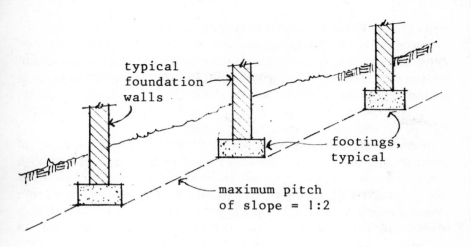

STEPPING OF ADJACENT WALL FOOTINGS

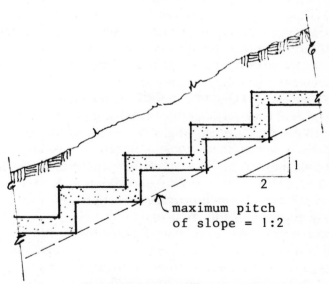

SIDE ELEVATION OF A CONTINUOUS
STEPPED FOOTING

STEPPED FOOTINGS

FIGURE 15

sand from the water-cement paste.

The strength required of concrete varies for different uses. For typical residential footings, concrete which has a minimum design compressive strength of 2500 lbs. per square inch (psi) is usually adequate. Higher concrete strengths are possible and are, indeed, required for many other applications. Concrete strength can reach as high as 8000 psi, and more.

(**CAUTION**) For consistency in quality control and reliability of mix strength, it is recommended that concrete be supplied by a ready-mix firm which specializes in this business, and delivers the pre-mixed material to the jobsite in the familiar concrete trucks, ready to pour. These firms furnish job delivery tickets which show the mix design strength, and can be relied on reasonably well to provide the specified mix strength and assure consistency between loads. You can request to be furnished copies of delivery tickets; and, if doubts arise concerning concrete quality and/or strength, insist on the taking of test cylinders to be tested by a laboratory, with the results furnished to you.

(**CAUTION**) Concrete must not be used if more than approximately 1½ hours have elapsed since the water was added to the mix. After that time, especially at higher air temperatures, the initial *set* or first hardening stage begins, and concrete should not then be placed or disturbed.

Hand or jobsite mixing of concrete should not be allowed. It is not unusual on very large projects requiring large quantities of concrete that a *batch plant* might be set up to handle materials and mix the concrete at the jobsite, in order to reduce costs and speed up delivery times. However, this would almost never be the case at a typical residential or sub-division site, and any other type of hand or small machine mixer would not normally produce the quality and consistency required.

Concrete which has been placed during very hot and/or dry weather conditions must be protected and **kept moist** after placement, to prevent premature evaporation of its water, which will result in impaired strength and hardness. Concrete placed in cold or freezing conditions must be kept **warmed** to at least 50 degrees F. by protection, including covering and heating if necessary. This process of protection is called *curing*. Curing must continue well beyond the *final set* period, which begins after about 8 hours.

FOUNDATION WALLS:

These are the below-grade walls which transmit to the footings the loads of the superstructure above. They are usually of cast-in-place concrete or of unit masonry such as concrete block. Since foundations extend below ground, they are subject to moisture and other factors which would deteriorate less durable materials. Codes stipulate minimum wall thicknesses for various foundation materials, which vary dependant on height of the wall, the material it's constructed of, whether or not it is reinforced, whether of solid or hollow units, and how many stories above it is supporting. The minimum thickness is usually 8 inches for cast-in-place concrete and for masonry.

If constructed of masonry, and if required to be reinforced (common in the west and in areas subject to seismic activity), vertical re-bars are installed at certain intervals, fully embedded in cement grout poured into the open cells, or spaces, of the masonry. See (Figure 11). Horizontal masonry reinforcement is also required, being placed in the mortar of the horizontal bed joints at frequent intervals, such as every 2nd or 3rd horizontal layer, or *course*. The vertical bars are not as frequent a requirement in the east and mid-west; however, the horizontal reinforcement is. See (Figures 11 and 12).

All unit masonry walls should have full mortar-filled horizontal and vertical joints. Vertical joints should not occur directly over one another in successive courses, but rather, should be staggered in alternate courses so that the blocks develop a natural bond (Figure 16).

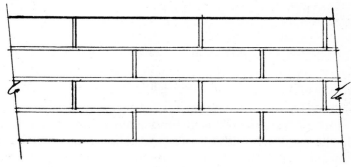

COMMON RUNNING BOND (ELEVATION)

FIGURE 16

Chapter 4: Footings and Foundations

Other than for the sake of appearance, the joints of masonry which is buried in the ground and exposed only to soil on both faces, need not receive special treatment. However, mortar joints of basement or cellar walls, and masonry exposed to inclement weather should be compacted and made more resistant to wind and water penetration by a process called *tooling*. Excess mortar is removed and the joint is shaped and smoothed into various profiles by special metal joint tools. See (Figure 17).

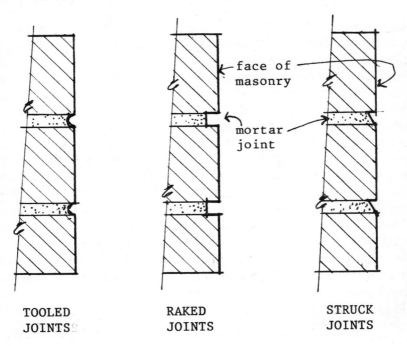

TOOLED JOINTS   RAKED JOINTS   STRUCK JOINTS

DETAILS OF MASONRY JOINTS

FIGURE 17

Before earth backfill is placed against any kind of basement wall, the walls should be braced either temporarily or by the permanent structural floor system, in order to prevent their movement, cracking or tipping-over from the forces exerted by the earth. See (Figure 2).

36   Chapter 4: *Footings and Foundations*

ANCHORAGE:

It is important that all adjacent parts of a building be adequately anchored or connected to each other for several reasons:
1) To resist up-lift due to wind action.
2) To resist lateral (horizontal) movement or displacement of components due to wind or earthquake forces.
3) To prevent displacement of components during the construction process.
4) To provide strong connections of dissimilar materials.

The first location requiring strong anchorage is between the foundations and the floor and/or wall systems which rest on the foundation. If wood-framed floor or wall systems are to be constructed, steel anchor bolts at least 1/2 inches in diameter should be set in the foundation so that the threaded ends rise up sufficiently to pass completely thru the wood *sill* or *plate* member which will be placed and anchored. The sill will then be secured with large steel washers and nuts drawn tight against the sill. If cast-in-place concrete is the foundation material, the bolts should be long enough to have at least 7 inches of embedment in the concrete, the bottom end of the bolt bent in an "L" shape for withdrawal resistance. If masonry units are the foundation material, bolts should be long enough to extend at least 3 courses deep into the block masonry, and have large plate washers on the bottom end which are embedded in a horizontal mortar joint. Cement grout is then filled into the open cores or cells of the masonry, encasing the bolts. Bolts should be located at approx. 4 ft. intervals, (some codes allow 8 ft. max), with a minimum of two bolts required for any single piece of plate or sill member being anchored. See (Figure 18).

DAMPPROOFING AND WATERPROOFING:

As a minimum precaution, all below-grade portions of basement walls should be dampproofed. This is done with special bituminous or asphalt materials of heavy consistency made especially for this purpose, applied by brush or trowel. 2 separate coats are recommended, so that skips or thin applications of a first coat are covered by the second. Carry dampproofing down and across the top of footings. See (Figures 12 and 19).

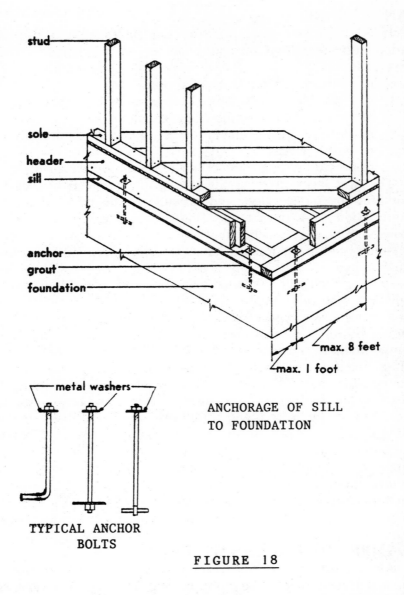

ANCHORAGE OF SILL
TO FOUNDATION

TYPICAL ANCHOR
BOLTS

FIGURE 18

As pointed out previously in CHAPTER I, concrete floor slabs resting on the ground may or may not be anchored to the foundation walls. In warm climates where footing depths are shallow, and especially where the exterior walls are wood framed, it is not uncommon for the floor slab to be poured integrally with

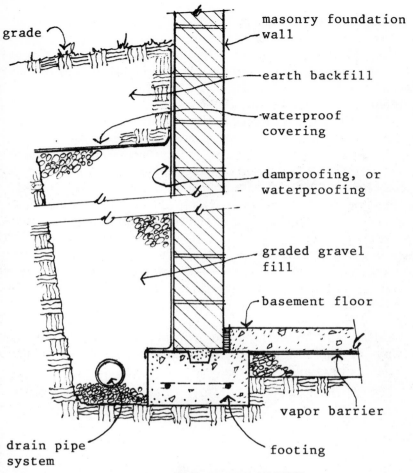

PIPED FOUNDATION DRAIN SYSTEM
FIGURE 19

the foundation (Figure 4). However, this is not usually recommended in colder climates, where an insulated separation is desirable (Figure 5).

If it is expected that wet soil conditions outside the foundation walls will prevail continuously, or be present in significant amount periodically or for extended periods of time, then a *membrane* type of waterproofing system should be installed. This utilizes a type of special waterproof sheeting—or membrane—which is adhered to the wall and then covered with a layer of hardboard or other protective material to keep backfill operations and materials from damaging it. One membrane system consists of alternate layers of asphalt or coal tar pitch and impregnated roofing felts, very similar to the make-up of built-up roofing. 2 plies should be the minimum application. This membrane should be carried down the foundation wall to the footing, across the top and down the outside face of the footing.

If severe standing ground water is expected, a foundation drain system may be required to relieve pressure and build-up of water. This is a pipe system which runs around the outsides of the footing of all excavated basement spaces. Special pipe is used which has small perforations, or holes, for the water to enter, or is special porous cement pipe which allows the ground water to seep directly thru the walls of the pipe. This piping system must be pitched to carry the collected water to dry wells or into a storm water drain system. Gravel should be backfilled around and over the top of this drain system, and the gravel covered with a waterproof felt or other covering which will prevent soil from seeping thru it into the gravel and then into and clogging the pipes. (Figure 19)

(**CAUTION**) Wood in direct contact with the earth should always **be avoided**, to prevent rot and infestations by termites, carpenter ants and other destructive actions of the soil.

CHAPTER V . . . FLOOR SYSTEMS

BASEMENT FLOORS:

Basement floors should be concrete having a minimum compressive strength of 2500 psi., 3½ to 4 inches thick, and be pitched to a previously installed drain or sump system. The floor slab should be placed over a vapor barrier of heavy (6 mil.) polyethelene sheet, or other suitable vapor-resistant material which will withstand construction traffic without puncturing. The vapor barrier is laid over a 4 inch layer of compacted gravel or crushed stone. The purpose of the vapor barrier is to prevent upward migration of moisture from the soil into the slab, resulting in dampness or wetness of the floor. Any joints or seams in the vapor barrier should be taped together with water resistant tape. The gravel layer should be clean, washed stone, free of soil and other fines which would negate its purpose of further preventing upward migration of moisture to the floor. The gravel must be well tamped to stabilize and compact it. See (Figure 19).

Basement floors should be separated from the surrounding foundation walls by a continuous pre-formed expansion joint material, a minimum of 1/2 inch thick.

If it is determined that a significant ground water problem exists, a *waterproof membrane* similar to the membrane discussed in CHAPTER IV for foundation walls, may be required under the floor slab. If installed, it is placed on top of the gravel layer in lieu of the vapor barrier, extend across the top of the footings, and be flashed (made watertight) to the membrane on the outside face of the foundation wall. Any penetrations thru the membrane such as floor drains, pipes, re-bar, etc., are potential sources of leaks, and must be very carefully flashed and sealed tight. See (Figure 20).

To prevent a build-up of hydrostatic pressure which can actually lift up floor slabs, a drain system beneath the floor **and** a foundation drain system as discussed in CHAPTER IV should be installed to collect the water and lead it to a method of disposal away from the building. In a location where ground water is truly a serious factor, it may be prudent to simply avoid the basement.

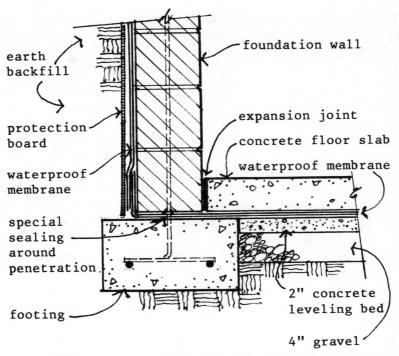

MEMBRANE WATERPROOFING AT BASEMENT FLOOR

FIGURE 20

CONCRETE FLOORS AT GRADE:

In warm regions of the country, grade level concrete floors can be 3½ to 4 inches thick, of 2500 psi. concrete placed over a 4 inch gravel, stone or A.B.C (aggregate base course) layer which is placed on the prepared and compacted soil base. Floors for garages or carports should never be less than 4 inches thick, and preferably be reinforced with 6" x 6", #10/10 welded wire steel fabric. These slabs may or may not be poured integrally with the stem, or foundation, wall, depending on the stem wall material and type of above-grade exterior wall it is supporting. See (Figures 4 and 11), and the discussion on ANCHORAGE in this CHAPTER.

In cold regions, grade level concrete floors should be isolated from the exterior foundation wall and be insulated to prevent significant heat loss at the floor perimeter. Such floors should also have the vapor barrier discussed previously for BASEMENT FLOORS. See (Figure 5).

Partitions which rest on concrete floor slabs and support floor and/or roof framing, require an equivalent footing under them, to uniformly transmit those loads to the soil. See (Figure 21) for acceptable alternatives of constructing the footing.

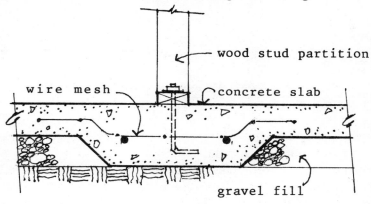

FOOTING POURED INTEGRALLY WITH SLAB

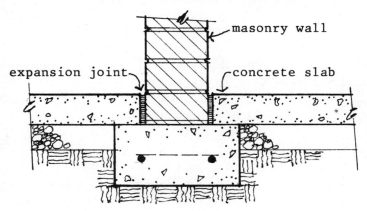

FOOTING POURED BY ITSELF

FOOTINGS FOR INTERIOR BEARING WALLS

FIGURE 21

Ideally, concrete floors should be installed before significant amounts of wood framing proceeds; especially before finish carpentry and cabinet work, so that the floor has a chance to set and dry out without its moisture being absorbed by the dry finish lumber, causing the wood to swell and warp.

The same concerns regarding *curing* of concrete as discussed in CHAPTER IV apply here, with even greater emphasis, because floors are relatively thin members with very large surface areas, and so are very subject to premature dry-out or to freezing.

There are available some very good chemical *hardners* which when applied (usually by spray) within specified times of the finishing of the concrete, will densify and harden the top surface much more than un-treated concrete would achieve. This can be a very desirable treatment for floors which will not receive tile, carpet or other covering, such as a garage or the basement floor. The hardner makes the floor more resistant to wear and surface damage from foot or vehicular traffic.

Large concrete floor surfaces will crack, due to shrinkage of the concrete as it sets and loses moisture. Certain locations of probable cracking can be anticipated, at which deliberate joints or breaks in the continuity of the concrete should be provided. These are called *control joints*. Control joints should be provided at: 1) Approx. 20 to 25 ft. intervals in large floor areas; 2) locations of major changes in the direction of the slabs; 3) Any point of significant *narrowing* or constriction in the shape of the slab.

Expansion joints should be provided in concrete floor slabs at their intersections with different materials, such as where basement slabs meet foundation walls, and around columns or piers which penetrate thru ground-supported slabs.

FRAMING LUMBER:

This is the *rough* lumber used to construct the structural skeleton, or frame, of a building. It is almost exclusively made from so-called "softwood" which is cut from evergreen trees such as fir, hemlock, spruce, etc. The lumber is sawn into dimensional sizes (2"x4", 2"x6", 2"x8", 4"x4", etc.), and either left with rough sawn faces, or machined smooth. It is then dried to lower the moisture content. Drying is done by either a forced hot air process in a *kiln*, or by simply being allowed to air-dry. Drying is important, because as the moisture content of lumber is lowered, the wood shrinks; and, if incorporated into a building in the non-dry, or *green*, state, would ultimately dry and shrink by itself, with all the unfortunate consequences of poor fits, loosened connections, etc. "Nail popping" is one common evidence of this, especially nails used to attach gypsum drywall.

Lumber is graded according to rules of several Agencies and Manufacturers Associations into varying grades of structural quality and appearance. Building Codes will stipulate the grades which can be used for framing purposes, and these will vary depending on the application; i.e., load bearing or non-load bearing, studs, joists, beams, etc. Generally, *construction* grade or better should be used for load carrying studs; No.2 or better for floor or roof joists, and No.1 or better for heavy beams, posts, and timbers. All of these grades allow certain size knots and imperfections such as splits, checks, etc.

The designated sizes of dimensional lumber are always nominal (theoretical), not actual, since the finishing and drying processes always reduces the cut member to its actual size. For example, a smooth-faced 2"x4" will actually measure 1½"x3½", a 2"x8" will measure 1½"x7½", and so on. Lengths, however, will be the actual specified dimensions (8ft., 10ft., 12ft., etc.).

WOOD FLOOR FRAMING:

The most typical wood frame configuration in use today for residential buildings, is called the *platform frame*. This simply means that each story is an entirely separate entity or platform the essence of which is shown in (Figure 22). Other configurations such as post and beam, braced frame and balloon frame do exist. However, it is not pertinent to the purposes of this book to discuss the particulars of each of those systems, since the basic points covered here will apply.

The repetitive floor support members in a wood frame building are called *joists*, selected for size based on how much distance they must span, their spacing, and the amount of load they must carry. Joists are normally 2" or 3" in thickness x the required depth such as 6", 8", 10" or 12". If the joists are undersized, or of a less acceptable grade of lumber than should have been used, the floor may feel *bouncy* or *springy* when walked on. This is a sign of excess deflection under load, and can be at the minimum very annoying, at the worst structurally dangerous. Joists are usually

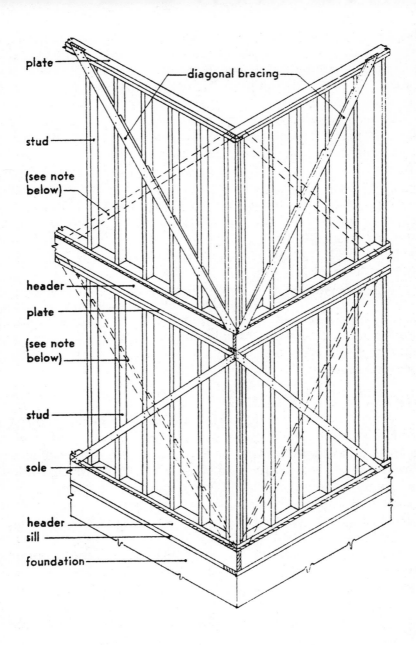

Note: Dotted lines indicate alternate locations of diagonal bracing

PLATFORM FRAMING, SHOWING DIAGONAL BRACING

FIGURE 22

spaced at 16" or 24" on centers, because these dimensions fit properly into the usual 4ft. or 8ft. modules of most materials (plywood, gypsum board, etc.)

Wood *sills*, or *plates*, which are bolted directly to the foundation as the base for the joists, as well as any other wood in direct contact with concrete or masonry, should be of one or two 2" members, preservative treated to protect against damage by moisture and insects. See (Figures 9, 12, and 18).

Joists should be supported by resting on a horizontal girder, plate, sill, or metal support connector at least 2 inches deep, preferably more. See (Figures 23 and 24). Joists should not be supported merely by end-nailing to the side of a member, with no bottom support.

All members in contact with each other must be nailed or spiked together, the quantity of nails being dictated by codes, size of members, loads, and good practice. A single nail per connection, even though it may be adequate in some instances to carry the particular load, is **never** acceptable. Nails driven into the sides or edges of lumber are much stronger and resistant to withdrawal than those driven into the ends. See (Figure 25).

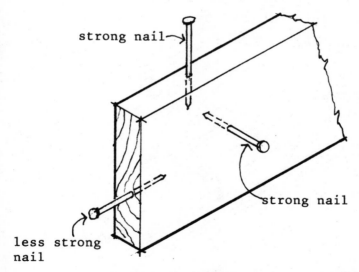

NAIL WITHDRAWAL RESISTANCE vs. FACES OF WOOD

FIGURE 25

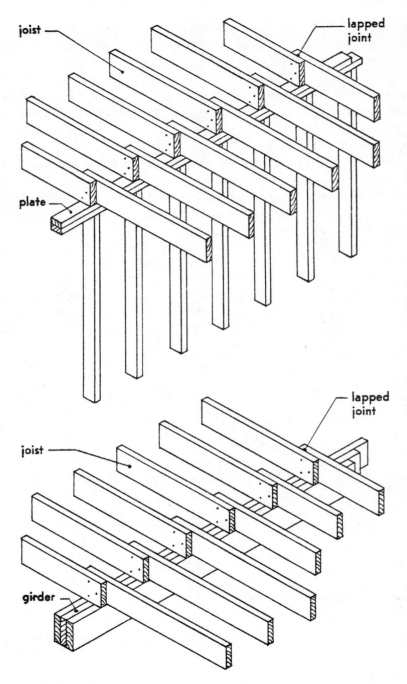

ACCEPTABLE JOIST BEARING CONDITIONS

FIGURE 23

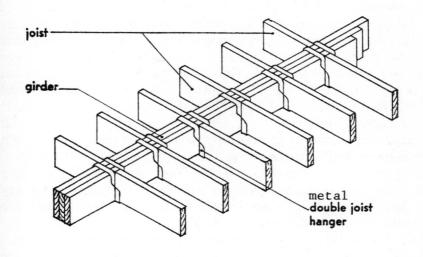

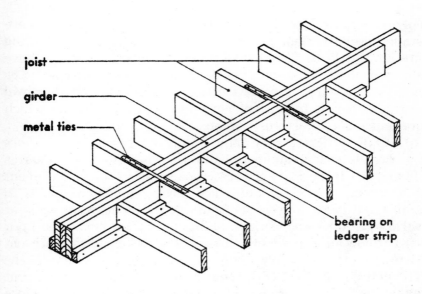

ACCEPTABLE JOIST BEARING CONDITIONS

FIGURE 24

FRAMING ANCHORS AND TIES:

All structural components of a building must be securely connected together to resist lateral and up-lift forces of wind, and to hold parts together under seismic (earthquake) action where this is a possible occurance. Anchorage to the foundation system is covered in CHAPTER IV, Article on ANCHORAGE. Connections are commonly achieved by mechanical fastners which include nails, screws, and bolts. A requirement of some local Codes—and **a good precaution in any building**—is the use of metal tie-down connectors, which are pre-formed shapes of galvanized steel, pre-punched or drilled for nails and other fasteners. These connectors are made in many gauges, shapes and sizes to suit a great variety of typical connection conditions. Local Codes will stipulate the frequency and locations for such tie-down connectors. If no Code requirement exists, use one at every other joist, stud, truss, etc. These connectors are nailed into the sides of members, and are always **in addition** to the normal nailing requirements for connection of frame members. See (Figure 26).

(**CAUTION**) Sub-flooring and/or underlayment material such as plywood, wafer board, or tongue-and-groove boards which are installed over the floor joists as a base for the finished flooring material, should be fastened to the joists with nails which have been specially coated with resin or are of other special types made for that purpose, to develop exceptionally high withdrawal resistance. This is to prevent noisy and squeaking floors which can occur due to nails working loose.

WOOD FLOOR systems should only be constructed *above* ground levels with an air space below the wood floor. See (Figures 9, 12 and 27). If the space is a crawl space rather than a basement the space **must** be ventilated. The most common means is to install fixed grills, louvers, or vents in the foundation walls at several locations around the building perimeter below the floor level, to allow air to move in and out of the space and thus prevent moisture build-up or dry rot from occuring. For example, the UNIFORM BUILDING CODE requires one square foot of net vent area for each 150 square feet of underfloor area, located as close to corners as possible, and arranged to provide cross ventilation. If there are signs, or concerns, about the dirt floor of the crawl space being wet or moist, it should be covered completely with a good heavy-duty vapor barrier. See (Figure 28).

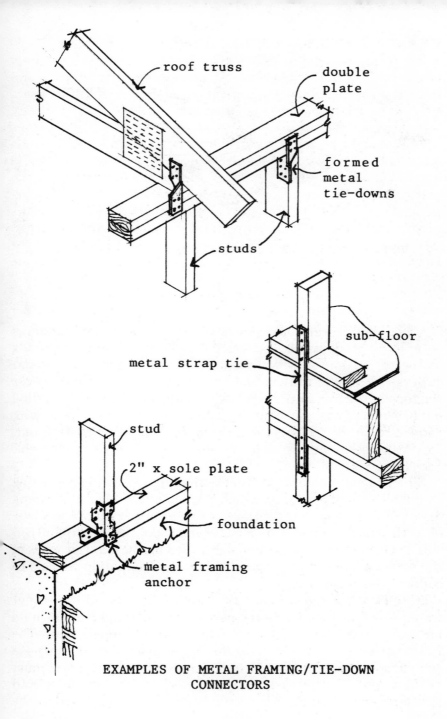

EXAMPLES OF METAL FRAMING/TIE-DOWN CONNECTORS

FIGURE 26

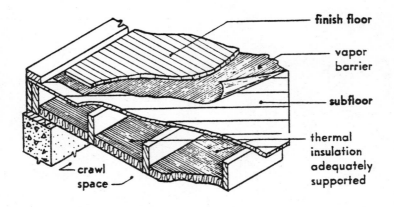

WOOD FRAME FLOOR CONSTRUCTION OVER CRAWL SPACE

FIGURE 27

A possible precaution against termite damage for buildings with above-grade wood floors is the inclusion of *termite shields* at the top of masonry walls on which wood floors, walls and girders rest. These are continuous metal sheetings which cover the tops of the foundation wall, with their outer edges extending out beyond the wall and bent down. The metal must be aluminum, copper, or protected steel to prevent corroding or rusting out. Joints and penetrations such as anchor bolts must be made tight. See (Figure 29).

All joists must be braced from deflecting sideways or twisting by installation of *bridging*, which is wood or metal cross-bracing installed in at least one continuous row at mid-span of the joists, or at the third points for unusually long joists. See (Figure 30).

If a veneer of masonry is being installed on the exterior walls of the house, a ledge must be provided on the foundation wall sill to support the masonry.

(**CAUTION**) *Flashing* must be installed at this sill to prevent water which will seep thru the masonry veneer and run down its backside, from damaging an adjacent wood-framed floor. The flashing can be one of several materials made for the purpose such as metal, fabric, plastic, metal-coated fabric, etc., and must be installed continuously along the sill, with all joints and penetrations sealed up. Also, *weep* holes must be provided at the

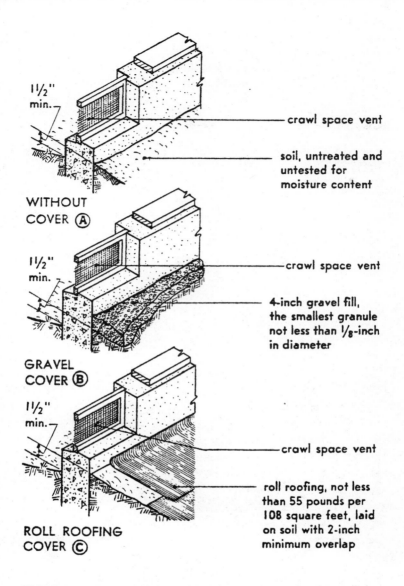

OPTIONAL TREATMENTS OF CRAWL SPACE DEPENDING ON MOISTURE CONDITIONS

FIGURE 28

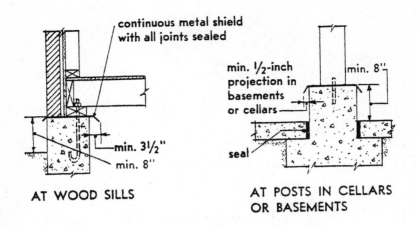
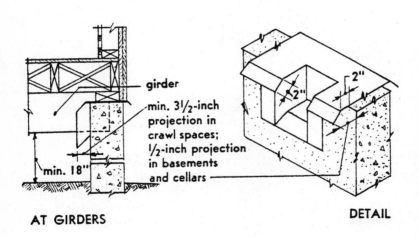

TERMITE SHIELD DETAILS

FIGURE 29

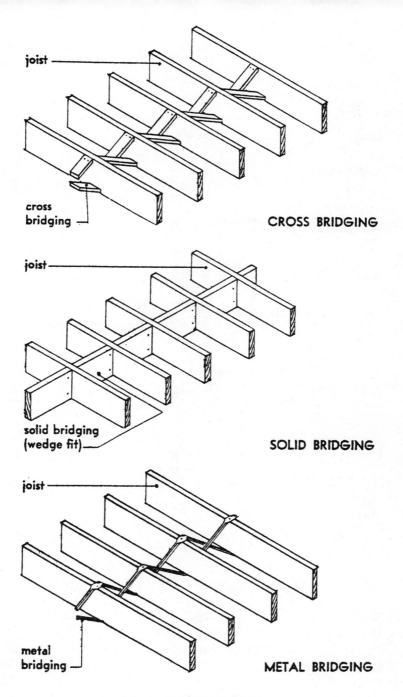

TYPES OF JOIST BRIDGING

FIGURE 30

base of the masonry veneer. These are simply small openings at periodic intervals to allow any accumulated water to seep (weep) out the face of the wall. See (Figure 31).

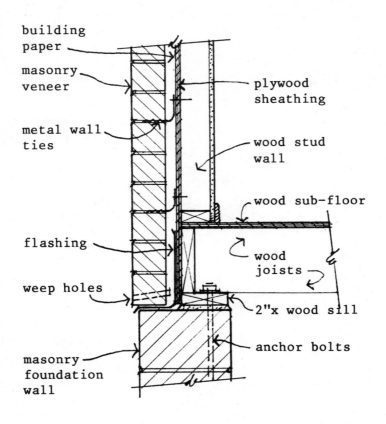

FLASHING AND WEEP HOLES AT SILL OF VENEER

FIGURE 31

Double floor joists should be provided under overhead partitions which run the same direction as the joists, if the partitions are non-bearing, or are only for the enclosure or division of spaces, not to support any floor or roof loads. Partitions which support other floor or roof framing should have load bearing partitions directly beneath them, or be located over main support beams or girders which have been designed to carry the extra load. See (Figure 32).

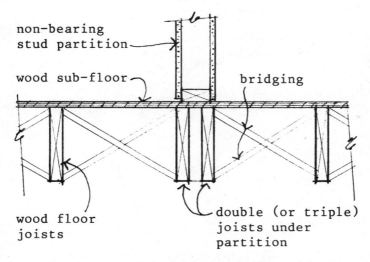

NON-BEARING PARTITION ON WOOD FLOOR FRAMING

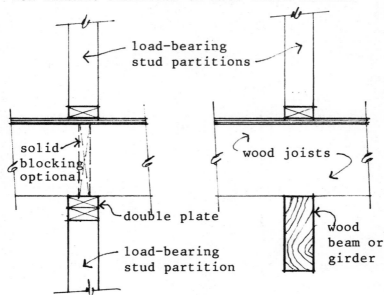

LOAD BEARING PARTITIONS ON WOOD FLOOR FRAMING

FIGURE 32

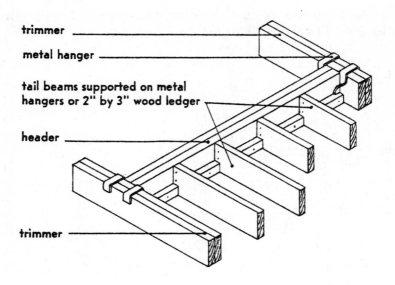

ISOMETRIC VIEW

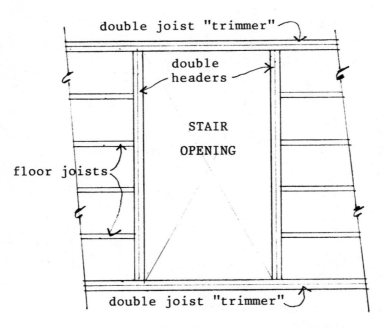

PLAN VIEW OF FRAMING FOR STAIR OPENING

FIGURE 33

Openings in wood framed floors for stairs or other purposes should have double joists all around the opening. See (Figure 33).

Wood members must not penetrate into a masonry chimney or fireplace, and should be placed at least 2 inches away from all such masonry surfaces. Similar, or greater, clearances are required around pre-fabricated metal fireplaces and flue assemblies. Be sure to check the fireplace manufacturers' written instructions for exact required clearances, since they can vary.

Upper floors of wood are framed the same as the first, except that the ends of the joists rest on the tops of wood stud partitions or on beams of wood or steel, instead of on the foundation sill. See (Figure 22).

(**CAUTION**) Large notches or holes cut in joists for the passage of large pipes or ducts **should not** be allowed, for such practices can significantly weaken or completely destroy their structural integrity. Small drilled holes for the passage of electrical cable are usually not detrimental, however, unless large quantities of them are drilled at the same locations.

CHAPTER VI ... EXTERIOR WALLS

Walls can best be described by the components which form their main structural elements. Exterior walls most commonly found in residential construction therefore, are wood frame, masonry, or wood frame with masonry veneer facing.

A *bearing wall* is one which supports loads from above, such as a floor, a ceiling or a roof. A *non-bearing* wall has no load resting on it. Interior non-bearing walls are more appropriately called *partitions*. Not all exterior walls are bearing walls, since floor or roof construction does not necessarily rest on all exterior walls of a building. Instead, load usually rests on some walls, while the remainder are of the same construction, but serve only as exterior enclosure.

WOOD FRAME:

Frame exterior walls consist of 2" x 4" or 2" x 6" vertical stud members placed at 16 inches or 24 inches on centers, set upon a base "sole plate" of matching dimensions. The sole plate attaches to the concrete or wood floor assembly acting both as a means of alignment of the studs as well as a fastening member. See (Figures 12, 18, and 22). Tops of the studs should be capped with two members, again called *plates*, for alignment and to form a resting surface for overhead construction. Open spaces between the studs must be insulated to improve the resistance of the wall to heat loss and heat gain. See the detailed discussion on INSULATION later in this Chapter.

Exterior faces of the studs should receive a *sheathing* which is a rigid covering of plywood, ¾" wood boards, or rigid insulation board, to which a layer of building paper and the exterior finish material is applied. See (Figures 34, 35 and 36). Exterior finish materials include wood or metal siding, plywood siding, stucco, masonry, etc. Interior finish materials such as gypsum board, wood paneling, lath and plaster, etc. are attached to the interior surfaces of the studs.

(**CAUTION**) In cold climates, a vapor resistive material, called a *vapor barrier*, should be installed on the interior of the wall framing prior to the interior finish materials, to prevent condensation of moisture within the wall.

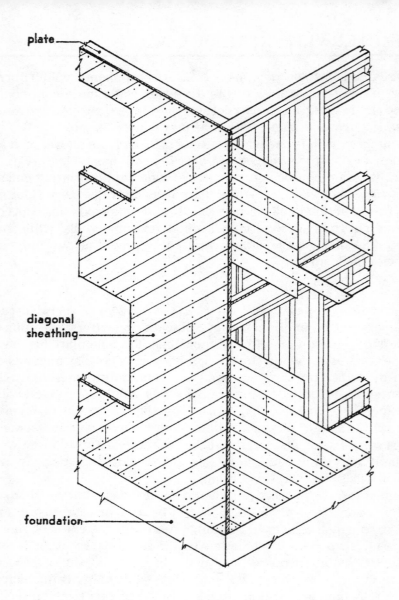

DIAGONAL WOOD SHEATHING ON WOOD FRAMED
EXTERIOR WALL

FIGURE 34

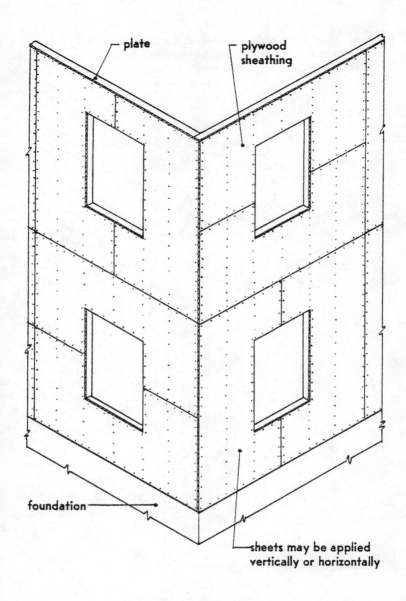

PLYWOOD SHEATHING ON EXTERIOR WALLS

FIGURE 35

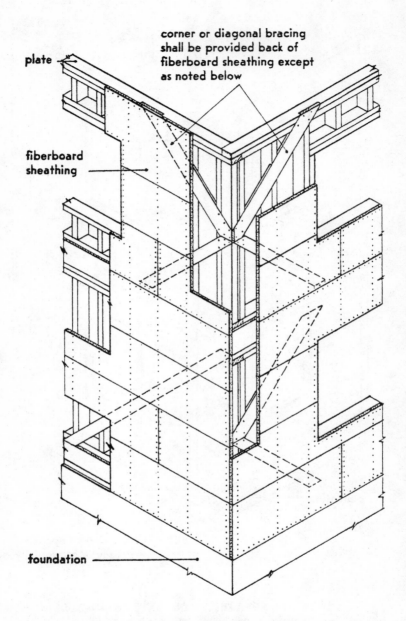

FIBERBOARD SHEATHING ON EXTERIOR WALLS

FIGURE 36

2"x6" studs are recommended in lieu of 2"x4", because of the increased insulation thickness which can and should be installed. 2 inch horizontal blocking the full width of the studs should be installed at approximately the vertical mid-point of stud walls for *firestopping*.

Stud walls must be braced in both directions to prevent their collapse from wind and other forces. Floors, roofs, and interior cross partitions provide the necessary bracing for forces which occur against the broad faces of the wall. However, to provide stiffness to resist collapse from the forces which occur along the long thin line of the wall, studs must have plywood sheathing (an **excellent** stiffner) or wood or metal diagonal bracing. Diagonal bracing members are cut into the faces of the studs, and extend from a high point at each corner, diagonally down, across and fastened to, several studs, terminating at and being fastened to the sill plate at the foundation. See (Figures 22 and 36).

(**CAUTION**) There is a cost savings trend in the warmer regions of the country where stucco is often used as the exterior finish material, for wood frame houses to be constructed using rigid expanded plastic board (such as Styrofoam), approx. 1 inch in thickness, as the exterior sheathing instead of a more dense or solid material. Wire "chicken mesh" is fastened over the plastic board, and the stucco then applied. While the foam plastic material does provide some slight additional insulating value, it has no nail or fastener holding capabilities, plus is very soft, easily broken and punctured. It would not take much force or effort, in the Author's opinion, to accidently (or deliberately) punch an opening thru a wall of this make-up. Security to unauthorized entry is therefore minimal.

Wood frame construction is very susceptible to damage from insects such as termites, water infiltration or leaks, and of course, is combustible. Wood can be specially treated to resist all of these destructive factors, but such treatment greatly increases the cost of the material, and so is not normally done.

MASONRY:

Masonry walls are constructed of brick or concrete block, the minimum allowable thickness for single story heights being usually 8 inches. Masonry may require vertical re-bar reinforcement depending on codes and location. If so, the bars are embedded in cement grout poured into the hollow cores of the masonry. All masonry should have horizontal reinforcement located in every 2nd or 3rd bed joints. See (Figure 11).

(**CAUTION**) Masonry is not a good insulating material, so in all climates some additional materials should be added to gain additional insulating qualities. A common method in use is to *furr*—or build out—the interior surfaces of the masonry with wood or metal strips which are fastened to the masonry, between which is installed insulation, and to the faces of which the interior finish is attached. The thicker the *furring* and insulation, the better the insulating value of the wall. Increasingly common today is to simply construct the equivalent of a 2"x3", 2"x4" or 2"x6" non-bearing stud wall alongside and against the interior face of the masonry, as interior furring, with full-depth insulation placed between the studs. See (Figure 37)..

In order to provide a proper support surface and nailing capabilities for floor or roof members which will rest on the masonry wall, a 2 inch thick treated wood member, or plate, is installed on top of the wall, anchored by steel anchor bolts exactly as described for masonry foundation walls in CHAPTER IV, under the discussion on ANCHORAGE. See (Figures 12 and 38). If the first floor is wood frame construction, flashing should be installed between the wood construction and the masonry wall, similar to that described for masonry veneer construction in CHAPTER V and in this Chapter. See (Figure 31).

As stated earlier, while masonry has a very low insulation "R" value, it nevertheless has a very high ability to slowly absorb significant amounts of heat; conversely, it is very slow to release that absorbed heat and cool down. These *thermal storage* and *thermal lag* characteristics can be used to advantage to help reduce dependency on mechanical heating and/or cooling systems. One way is to attach the insulation and a protective finish to the exterior face of masonry exterior walls, leaving interior faces exposed. By so doing, the masonry is not subject to wide swings of exterior heat build-up or heat loss, while at the same time its thermal lag aspect tends to even out and stabilize interior temperatures.

In houses designed to take advantage of solar heat, masonry partitions and walls can be strategically located so as to be heated by direct sunlight, with the subsequent slow release of

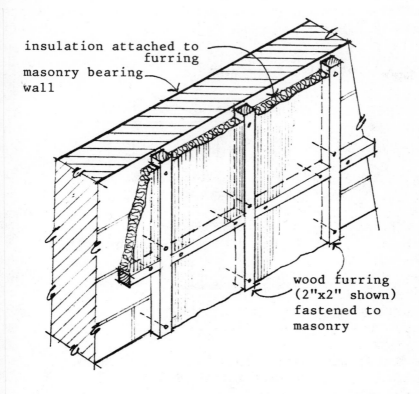

FURRING AND INSULATION ON INSIDE FACE OF WALL

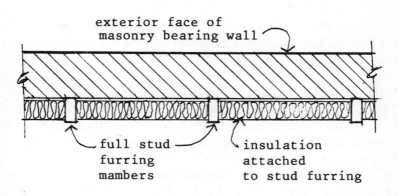

PLAN VIEW SHOWING FULL STUD FURRING

FIGURE 37

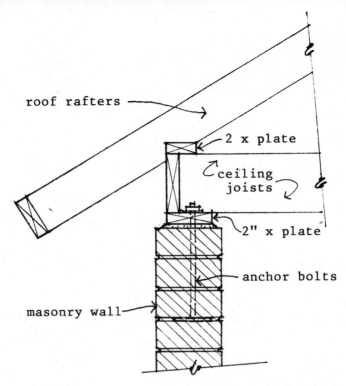

ROOF AND CEILING CONSTRUCTION ATTACHMENT
TO MASONRY BEARING WALL

FIGURE 38

that heat being a major factor in the heating of the interior spaces after the sun has set. Much research and experimentation on this phenomonen has been done in the southwestern United States.

WOOD FRAME WITH MASONRY VENEER:

The structural load-carrying portion of this assembly is a wood framed stud wall exactly as described earlier. The differences in the veneer wall are that the exterior finish is a layer of brick, block or other masonry which is non-structural and non-bearing, but is attached to the back-up stud wall for its support. The sheathing material on the outside face of studs must be of plywood or wood boards which have nail holding capability, so that the veneer can be periodically fastened—or tied—to the back-up wall. This is necessary because the veneer is relatively thin (usually 4 inches or less in thickness), and does not have the stability to stand alone without back-up support. The usual method of attachment is by galvanized metal strips called *wall ties*, which are installed in the joints of the masonry as it is being laid up. The wall ties are then bent up and nailed to the sheathing, preferably also into the studs. Because water can, and will, permeate thru most masonry veneer, there should be an air space separation between it and the wood sheathing; and, flashing plus weep holes should be provided at the base sill, as explained in CHAPTER V. See (Figure 31).

INSULATION:

Heat creates molecular activity in materials; the higher the amount of heat, the higher the activity. Molecules of higher activity transmit part of their energy to those of lower activity. Therefore, heat **always** flows from the hottest materials to the colder. This is the principle of *thermal conduction*.

Each material has a certain amount of resistance to the conduction—or flow—of heat thru it; some, such as metals, are very low in resistance, and some such as insulating materials have very high resistance to heat flow.

The resistance to heat flow for a **1 inch thick** one square foot section of any material is expressed as the coefficient 'k'. 'k' = the rate of heat flow (BTU's) per degree of temperature, per hour, per square foot, per inch of thickness.

The total thermal resistance of one square foot of a particular material of a **fixed thickness** is called its Resistivity 'R'. 'R' = the resistance to heat flow (BTU'S) per degree, per hour, per square foot for the particular thickness. The greater the 'R' number of a material, the greater the resistance to heat flow, and therefore the greater the insulation value.

The summation of all the individual 'R' factors for all materials in a wall, floor or roof equals the total thermal Resistance of the assembly.

$$\text{'R'}_{wall} = R_1 + R_2 + R_3 + \text{(etc.)}$$

The overall expression of heat transmission thru walls, roofs, floors, windows, etc. is expressed as 'U'. The 'U' value is equal to one (1) divided by the sum of all the individual 'R' values of the materials in the assembly.

$$\text{therefore, U} = \frac{1}{R_1 + R_2 + R_3 + R_4 + \text{(etc.)}}$$

Since 'U' is the reciprocal of 'R', the **lower** the number for 'U' the **better the insulation** value of the assembly. Thus, a wall having a 'U' = 0.1 is much better at resisting the flow of heat thru it than one with 'U' = 1.0, by a factor of 10. The opposite would be true for the total Resistivity 'R' of the wall.

As the above would suggest, 6 inches of a certain kind of insulation has greater resistance 'R' to heat flow than 4 inches of the same material. As proof of this, following are 'R' values listed in 1987 by the Certaineed Corp. for various thicknesses of their *Fiber Glass Building Insulation Blankets*:

Blanket Thickness	R value
2½ inches	8
3½ inches	11 and 13
6¼ inches	19
6½ inches	22
10 inches	30
12 inches	38

There are several sets of energy standards in effect for the country, which set minimum levels of total resistivity 'R' for residential floors, sidewalls and ceiling. The recommendations vary depending on the location or 'Zone' of the country in which the residence is located. In some areas, heat loss from the building due to cold exterior temperatures and wind chill factors is the main determinant; while, in other areas the need to cool or *air-condition* because of the generally higher prevailing outdoor temperatures controls; and, in still others, it is a combination of both. In general today the absolute minimum 'R' values in the most mild Zones are as follows:

Ceiling, or Attic Floor	Sidewalls	Floors
R-19	R-12	R-11

All other Zones increase from these values. See **APPENDIX 'C'**.

There are many types of materials used to make insulating products. All have different 'k' values per inch, and therefore different 'R' values for various thicknesses. Some are made in flexible blanket form such as the familiar fiberglass insulations, others are of pre-formed rigid board material which includes Styrofoam, urethane, wood fiber, etc. Still others can be custom job-formed such as spray-on urethane. Each insulation is made for a specific set of purposes and applications, which should be investigated before use.

CHAPTER VII . . . WINDOWS, GLASS, DOORS AND HARDWARE

WINDOWS:

Windows are generally classified by their type of operation. There are fixed (do not open), double-hung (slide up and down), horizontal sliding (well named), casement (side hinged, like a door), and projecting (swing out or in, on horizontal pivots at top or bottom). See (Figure 39).

Windows are composed of the **frame**, which connects to the building wall, and the **sash**, which is the operable or movable portion.

The other major variables in the windows are the materials the frames and sash are made of, their sizes, and kind of glass they can hold.

Selection as to type of operation is mostly a matter of deciding how much of the window you wish to have open to outside ventilation, and how much ability you want to change the direction of the incoming air. For example, the double-hung and the horizontal sliding windows can open a maximum of 50% of the overall window size (usually somewhat less), but each affords no ability to control the direction of any incoming air. On the other hand, casement and projected windows can open almost 100% and do provide some amount of control on air movement patterns. This is because, like a damper or control vane, they angle the air in varying degrees, depending on the position of the sash.

Selection of the materials of which the frame and sash are made is dependant on several factors. Metal windows are normally made of steel, which must be protected from rust; or, of aluminum which is lightweight, however less rigid, but do not rust nor require painting to survive. Aluminum windows are available with various factory finishes and coatings which do enhance their appearance as well as eliminate grey oxidation, or discoloring, which eventually takes place with plain unprotected aluminum. The advantages of metal windows are the crisp, thin frame profiles which can be a design or visual asset, and their resistance to rot and decay compared to a window made of wood. They may also be slightly cheaper than wood windows, but this can vary with locale.

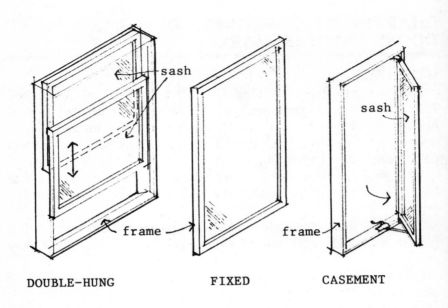
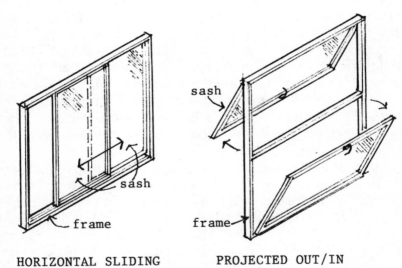

TYPES OF WINDOWS

FIGURE 39

Wood windows, and windows of wood encased in a plastic jacket are available for residential construction. These have assets and liabilities which are directly contrary to those of metals; however, they are usually somewhat more costly and are always more bulky in lines and appearance. If of un-protected, or plastic-encased wood, they require more care and maintenance, such as painting.

(**CAUTION**) In very cold or very hot climates, metal windows have a very common defect, which is their ability to easily conduct heat. In those climates, the frame and sash parts transmit heat readily to either the outside or inside, thus contributing to the mechanical equipment load of the building. Also, depending on the level of humidity in the air on the warm side of the window, moisture condensation can easily occur with its annoying and damaging effects. Metal windows are made which have special split construction to break up the conduction path thru the metal and thus minimize these problems, but they are usually precluded from residential construction due to their higher cost.

Another cost factor, although admittedly not a major one, in the selection of windows can be whether or not special hardware is required to make the window function. For example, almost no special hardware is required for a horizontal sliding aluminum window, as opposed to special offset hinges, geared operator, and latch locks which are all necessary for casement type windows. In some respects, however, the old adage *you get what you pay for* applies.

GLASS:

Beyond the concerns for cost, type of operation, condensation/conduction effects of the frame and sash, and the appearance, the **most important window decisions** to be made are the **quantity of glass** to be installed, and its **thermal, or insulating qualities**. The insulation quality of clear sheet glass is directly affected by the number of parallel sheets, or panes, of glass that are used. Quite simply, the more sheets with thin air space separations between each, the better the insulation value. It is very difficult to envision a location or a situation wherein the installation of at least double-pane glass would not be beneficial, whether in the form of manufactured sealed *insulating* glass (Thermopane,

Twindow, etc.,), or simply by the use of removable *storm windows* commonly found in the East and Mid-West. The reason for this recommendation is based on the fact that a single pane of glass in a vertical position has a heat transmission (U) value of **1.13**, which is **extremely high**. Remember from the discussion on INSULATION, the **lower** the 'U' value, the better. Two parallel panes of glass reduce the 'U' value to approx. **0.55**, while three parallel panes reduce it still further to approx. **0.36**. (Remember, there must be a slight air space between each pane; the actual values of 'U' vary slightly dependent on the thickness of the air space between each pane). Now, compare these 'U' values to that of **0.08** for the solid sidewall construction having a Resistance 'R'=12 required in the most liberal Zone of the country (Appendix 'C'); and, you will readily conclude that no matter what is installed for window glass, it will result in much less insulating value than any other portions of the building.

Glasses other than clear are available which can limit the amount of heat passing thru due to direct solar exposure. These include reflective-coated and color-tinted, available in single sheet glass as well as sealed insulating glass. Although more expensive, these specialty glasses should be considered for large areas of direct solar-exposed glass in the warmer southern and southwestern locations where solar heat gain is a major factor in the cooling/air conditioning load of a building. The measure of effectiveness against the passage of heat thru special glasses is the *shading coefficient*. For example, the coefficient for clear untreated glass is approx. 0.95, whereas reflective coated glass can be as low as 0.23.

A supplemental benefit gained by installing multi-pane windows lies in the fact they normally will not form moisture condensation on inner or outer faces during extremes of temperature and humidity.

The discussion above on the relatively poor insulating quality of glass, underscores the need to be concerned with the **quantity** of glass to be installed. The greater the quantity of glass area compared to the remaining area of insulated solid wall, the poorer the overall heat loss/heat gain characteristics of the building. You must therefore, carefully balance and decide on the optimum amount and type of glass area which can be allowed and afforded, vs. the need to conserve energy, preserve interior comfort and achieve lower costs of heating and cooling.

AIR INFILTRATION AND WEATHERSTRIPPING:
The final point of concern regarding windows is the need for a good *fit* of the window parts, plus the need for weatherstripping. Other than heat loss or gain thru the glass areas of windows or doors—and even more significant than conduction of heat thru metal frames—is the resistance of the window to *infiltration* of air thru loose fitting parts. Substantial amounts of air can be lost or gained thru poorly fitted windows, a factor which is taken into account in the selection of heating and air conditioning equipment capacity. All windows which have moving sash should have at least single, and preferable double layers of weatherstripping. This is merely flexible gasketing-type material which, like the rubber seal material around the door and windows of automobiles, maintains positive contact to moving parts and minimizes leaks of air and water. Ask your builder to show you the extent of these features on the windows being proposed in your building.

The **major weakness points** in today's construction as far as heat loss or gain are concerned, are the windows and doors. This is directly attributable to the quantity of glass, the number of panes of glass, and the quality of fit of the moving parts.

DOORS:
Doors are classified by their suitability for exterior or interior use, by their details of construction, and by their method of operation, i.e., swing-hinged, sliding, pocket, etc.

DOOR CONSTRUCTION AND TYPES:
Only certain doors are suitable for use in exterior walls of a building, where they will be exposed to the weather. Exterior doors must be assembled with water-resistant glue, or the door will gradually come apart as moisture penetrates and dissolves glue in the joints. Exterior doors can be used for interior purposes, but not the other way around.

PANEL doors are those which have exposed solid side, top, bottom and intermediate rails (usually of wood), which are the main structural parts of the door. The spaces between the horizontal rails and vertical stiles may be plywood, solid wood, glass, plastic or grilles or louvers for ventilation purposes. Panel doors are generally more decorative in appearance because of the shapes and moulding effects used. Good quality Douglas Fir or other appearance grade wood is usually used in their construction, which allows for natural or stained finishing as well as paint finishes. See (Figure 40).

FLUSH doors have flat faces, usually both sides of the door, which may be plywood, hardboard, or metal. Flush doors are either *solid core* or *hollow core*. Solid core doors have an interior construction of solid wood pieces glued together, or the core may be of solid particle board or a fire-resistant mineral material. The top, bottom and edges are usually continuous pieces of solid wood. See (Figure 40). Flush exterior doors are usually solid core, as are other special purpose doors in commercial and institutional buildings, because of their increased durability and strength. Flush doors may have various hardwood veneer facings, suitable for natural or stained finishes, or be of less exotic, less costly materials more suitable for painting.

Hollow core doors differ from solid core in that the inner core of the door consists of an open grid or lattice work of crossed wood slats with considerable void spaces between them. These doors are much lighter in weight and are more generally made for interior use only. See (Figure 40).

BATTEN doors consist of solid wood boards fastened together side-by-side, with another layer of wood members, called *ledgers*, attached straight across or angled across the faces of the first board layer. See (Figure 40).

DOOR SIZES:

Doors are made in thickness of 1⅛ inches, 1⅜ inches, 1¾ inches and 2¼ inches (rare). Exterior doors should be not less than 1¾ inches thick, because of their larger overall sizes, and to accommodate the heavier-duty hardware required. 1⅜ inch thick interior doors are common, and should be the **minimum** thickness for hinged doors up to 2ft-8inches wide; over 2ft.-8inches, doors should be 1¾ inches thick. 1⅛ inch doors can be used for narrower doors and less critical applications such as bi-fold closet doors. Stock door heights are 6ft-8inch, and 7ft-0inch. 6ft-8inch is suitable and most usually used for interior doors. Stock widths vary in 2 inch increments from about 1ft-6inch up thru 3ft-0inch. Door widths should vary depending on the

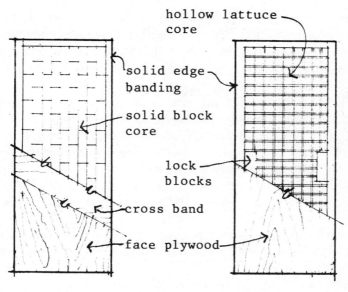

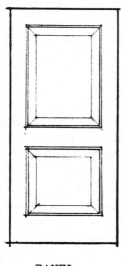
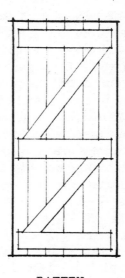

TYPES OF DOORS

FIGURE 40

location and purpose of the door. At least one 3ft-0inch wide exterior door should be provided in every house to allow adequate in-and-out movement of furniture and equipment. Two such doors are better, at obvious major entry/exit locations (front and rear, front and side, etc.).

(**CAUTION**) Occasionally, interior doors are undersized in width, making it difficult to move furniture and belongings in and out of spaces. If there are doubts about the adequacy of a particular door size, increase to the next standard width, rather than accept the constraints of the smaller opening. The increment of cost is very minimal. Also, if there are handicapped or disabled persons in the household, certain minimum door openings will be required to accommodate their needs, such as 32 inches clear (minimum) for wheelchair access, etc. Any door less than 32 inches will present difficulties in furniture movement thru it.

METHODS OF OPERATION OF DOORS:

HORIZONTAL SLIDING doors suspended by metal or nylon rollers from overhead metal tracks are often used for interior closets and storage spaces. By-passing sliders (Figure 41) can provide from 1/3 to 1/2 of the total opening as actual access opening; whereas, if all doors stack behind each other, 2/3 of the opening can be accessible.

POCKET doors are horizontal sliders which slide into a pocket created within the adjacent partition, and thereby are completely hidden and out of the way when open. See (Figure 42). A pocket door can be very useful in a location where closure of the door is only infrequently required; whereas, if hinged, the door when standing open would be an inconvenience or an obstruction. Pocket doors are not recommended for high-traffic locations where the door must be frequently opened and closed, because their operation is time consuming and clumsy.

SLIDING EXTERIOR doors (commonly called *patio* or *Arcadia* [which is actually a trade name] doors) are simply a form of by-passing horizontal slider which are specially constructed and weatherstripped for exterior wall usage. Here again, quality of construction, good fit of parts, lots of weatherstripping, and use of multi-pane insulating glass are important. They are available in wood, steel, and aluminum.

SWINGING HINGED doors are the most common operation

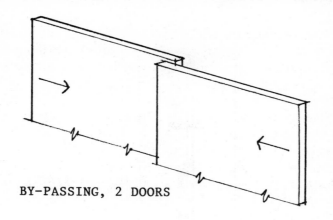

BY-PASSING, 2 DOORS

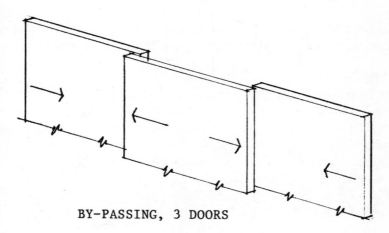

BY-PASSING, 3 DOORS

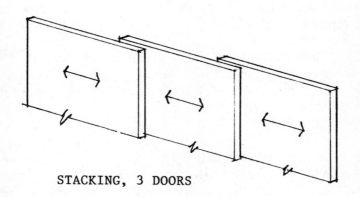

STACKING, 3 DOORS

TYPES OF BY-PASSING DOORS

FIGURE 41

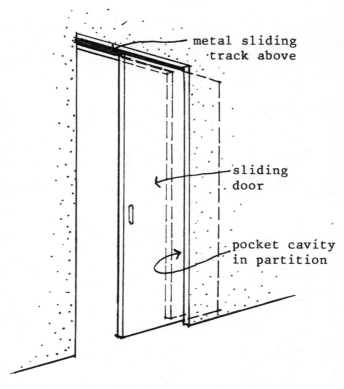

POCKET DOOR

FIGURE 42

found. These doors pivot on hinges mounted to the frame, close against *stops* along all sides of the frame, and are held closed by a latch or lock mechanism.

ROLL-UP doors consist of many small interlocking sections, usually steel or aluminum, which open by rising upward while coiling around a drum assembly at the head—or top—of the door opening. These are rarely used in residential construction.

SECTIONAL OVERHEAD doors are often used for large garage doors. They consist of horizontal sections full width of the opening but short in individual height, made of wood or metal, flush or paneled, connected together by hinges. The sections are guided by ball bearing rollers in side metal tracks and counterbalanced by various kinds of spring mechanisms, so that they rise up and lay overhead, above and out of the way of the door opening.

The DOOR FRAME is the surrounding unit to which the door is hung and against which it closes. The frame is a finished member which also serves to cap the rough wall opening. Frames are made of wood or metal. Wood interior frames are a minimum of 3/4 inches thick, if the *stop* or piece to which the door closes against is nailed onto the frame. Exterior wood frames should be thicker (a min. before milling of 1¼ inches), with the *stop* being an integral and notched-out part of the frame rather than nailed on the surface, See (Figure 43). This is for reasons of security.

HARDWARE:

All hinged doors should have a minimum of three (3) butt hinges, in order to resist warpage, sag and poor fit. *Butt* means that the hinge is mortised, or housed, into the edge of the door, rather than fastened to its surface. Batten doors, however, usually require surfaced mounted hinges.

(**CAUTION**) If for some reason an exterior hinged door must swing out, with the hinges therefore exposed to the outside, the *pin* of the hinge which holds the two hinge parts together, and about which they pivot, should be a special self-locking pin so that the door cannot be removed from its hinges when closed.

LATCH sets provide a knob or lever grip to operate the door, and a mechanism to hold doors in the closed position via a *latch* which engages the frame. A LOCK SET is similar to a latch set, but provides the ability to lock and unlock the door with a key. Locksets should have multi-tumbler cylinder keying, not be the old fashioned *bit* key, both for proper security and for compatability with today's technology. Latch and lock sets are either *mortised*—or housed into—the stile edge of a door, or are of the tubular bored-in type, which is mounted into a hole bored thru the faces of the door. The tubular type are much more common today and much less costly to install. See (Figure 44).

There are many other items of hardware which may be useful and proper depending on the location, need, etc. These include door closers, dead locks, dead bolts, kick plates, etc. They are considered beyond the scope and purpose of this book. Consult your Builder or hardware Supplier.

There is a wide variety of finishes and designs available for hardware.

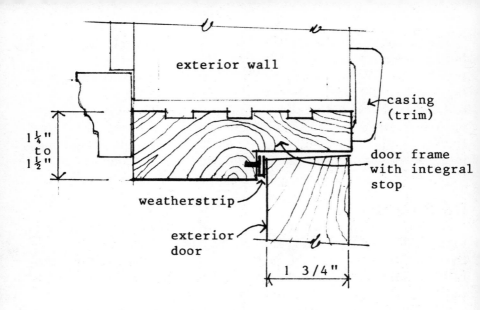

SECTION THRU EXTERIOR DOOR FRAME

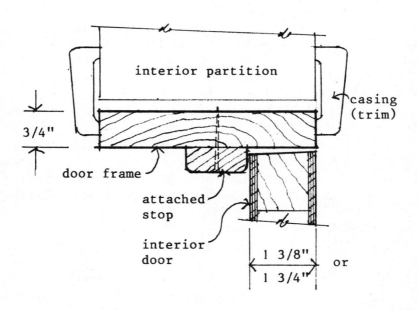

SECTION THRU INTERIOR DOOR FRAME

FIGURE 43

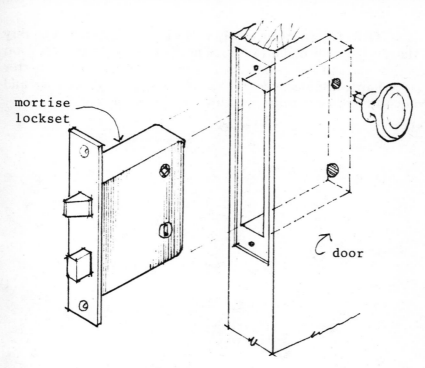

MORTISE LOCK

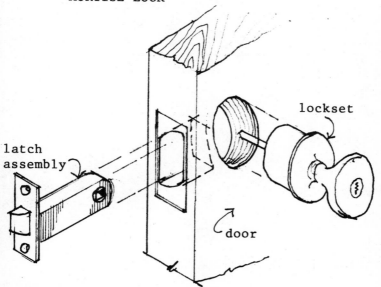

TUBULAR, OR BORED-IN LOCKSET

FIGURE 44

(**CAUTION**) More importantly, there is a wide range of quality in the mechanisms and make-up of hardware. Generally, it is good practice and value to spend a few extra dollars to buy better quality hardware, because its durability, improved security and trouble-free use will prove worthwhile in the long run.

CHAPTER VIII... THE ROOF STRUCTURE

SHAPES OF ROOFS:

Roofs can be constructed in many shapes. The *flat* roof is a single plane, almost—but not quite—level. A *shed* roof is a flat roof, but one which has definite pitch or slope. A *gable* roof consists of two inclined planes which meet at a peak over the center line of the house. The peak of the gable is called the *ridge*. A *hip* roof is one which slopes down in four inclined planes. A gable roof may have hip ends. A *mansard* roof is one which has sloped planes around its exterior perimeter, but is flat in the upper center portions. See (Figure 45).

(**CAUTION**) All roofs should have some pitch, so as to shed water. Roofs must not be constructed dead level, or so as to hold or *pond* water. Many years of sad experience have demonstrated that there is no absolutely watertight roofing material or system that will be able to remain leak free with standing water on it. Manufacturers of roofing materials will not guarantee nor stand behind their materials or systems if the roof does not have at least minimal pitch. The actual amount of pitch required varies for different roofing materials; some requiring as little as 1/4 inch per foot, others requiring significantly more to be trouble-free. See CHAPTER IX... ROOFING, for more discussion on this.

PARAPETS:

A parapet is a configuration in which the roof terminates into an upward extension of the exterior wall. Parapets are built at locations where it is desirable to not have roof run-off water, or for other reasons related to the design of the building. See (Figure 46).

(**CAUTION**) Parapets require special flashing treatment to make the intersection of the roof plane and the vertical wall watertight. See (Figure 47). Also, roofs which pitch water **towards** a parapet should be avoided because this traps the water in a valley, tending to cause standing water conditions with subsequent leaks. Parapets also require special waterproof materials to cap their top. Parapets, therefore, are not altogether desirable, and if used, require very special care in the initial installation of their flashings as well as frequent maintenance attention.

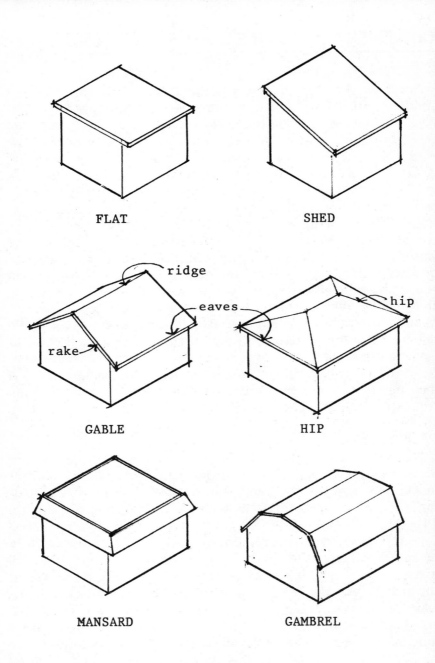

ROOF SHAPES

FIGURE 45

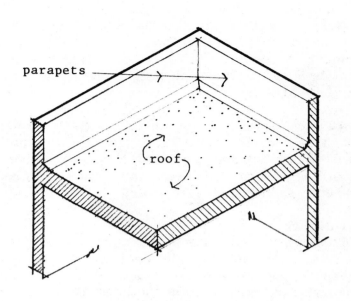

VIEW OF ROOF MEETING PARAPET

FIGURE 46

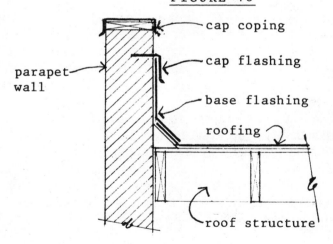

DETAIL OF PARAPET-TO-ROOF FLASHING

FIGURE 47

Wherever possible, roof planes should extend beyond the exterior faces of the building. This directs water away from the building, as well as provides shade from the affects of the sun. The amount of *overhang* of the roof is determined by design considerations, building orientation, climate, and the need for exposure to, or protection from, solar radiation.

All pitched roof structures inherently create structural forces which are constantly at work trying to cause the plane of the roof to move outward and downward, towards the direction of the pitch. These forces must be resisted by bracing, ties, collar beams, and other supports. See (Figure 48).

TRUSSES:

Trusses are a structural system quite commonly used in roofs of all types of buildings, including residences. A truss is a geometric configuration of structural members arranged in triangular patterns, wherein the intersection of each member requires a special connection. Trusses can be made to span long distances free of intermediate support other than at their ends; and, utilize members which are relatively small in size. Trusses made of wood are relatively inexpensive. They are a viable alternative to span longer distances which would be excessive for normal solid members, or where interior intermediate supports are not possible nor desirable. They also form large voids for the easy passage of ducts, pipes, etc. See (Figure 49) for some truss configurations.

(**CAUTION**) The chief concern regarding the use of trusses, is that their design must be done by qualified professionals, and their construction—especially the joint connections—be executed exactly as per the design. If there is any doubt about the design or fabrication of a truss proposed for your building, insist on the calculations, have them checked by another professional, and require a sample truss to be load-tested.

VENTILATION:

Any parts of a roof system which will result in trapped or closed-in *dead air* spaces, such as attics, **must** be ventilated. The reasons for ventilation are two-fold:

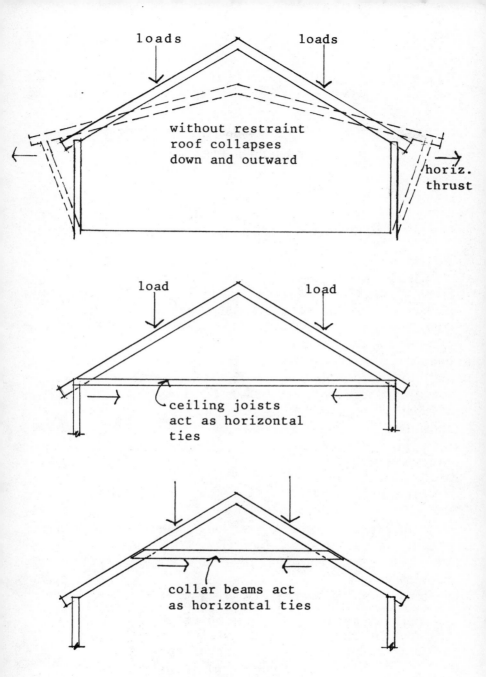

STRUCTURAL FORCES IN PITCHED ROOFS

FIGURE 48

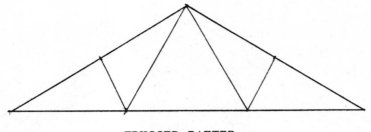
TRUSSED RAFTER

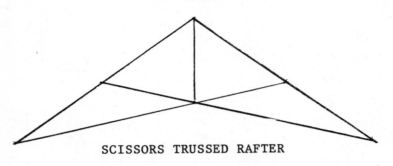
SCISSORS TRUSSED RAFTER

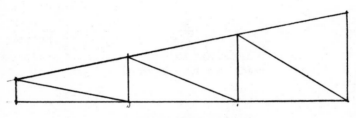
PITCHED PRATT TRUSS

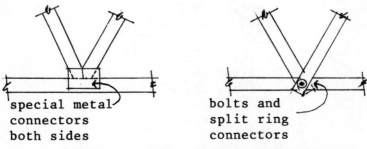
TYPES OF TRUSS JOINT CONNECTIONS

FIGURE 49

1) To allow hot air which builds up in trapped spaces to escape to the outside, thereby diminishing the effects of that heat on the interior of the building; and

2) To relieve the build-up of moisture which occurs in trapped air spaces due to the migration of humidity (water vapor present in heated air) into those spaces, and then condensing out of the air as liquid moisture. This moisture can wet and soak insulation, thereby significantly diminishing its effectiveness. It can also cause deterioration of framing members, rusting of nails and fasteners, and stain interior surfaces such as ceilings and walls.

(**CAUTION**) Ventilation should be provided around the perimeter of dead air spaces at both low points and high points. This allows natural convection currents as well as wind to move air thru the dead space. Vents can be fixed openings provided in the bottoms—or soffits—of roof overhangs, be special continuous vents at ridges, louvers high in side walls, and wind or electric-powered moving ventilators. There are standards and formulas provided by codes, and by the FHA for the amount of open ventilation area required for various kinds of spaces and conditions. See (Figure 50) for various ventilation schemes.

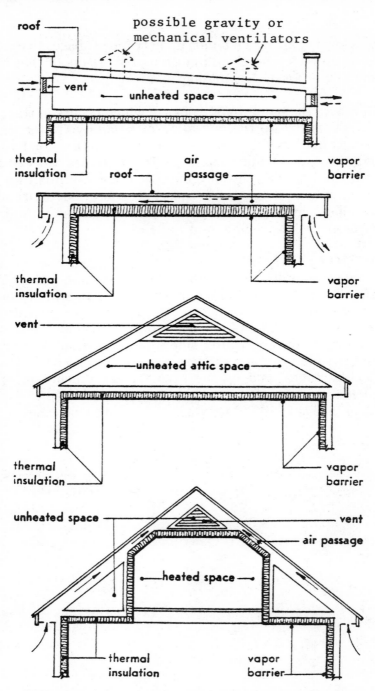

VENTILATION OF ROOF AND ATTIC AIR SPACES

FIGURE 50

CHAPTER IX... ROOFING AND FLASHINGS

Of all the materials and components involved in a building, the one which consistently causes the greatest number of problems, building damage and inconvenience, is the roofing system.

The roof system consists of the structural support system for the roof as discussed in CHAPTER VIII; plus the deck; and the materials, including flashings, which together comprise the *roofing*. This CHAPTER will discuss decks and roofing.

BASIC CRITERIA AFFECTING SELECTION OF ROOFING:

The major factors which influence and control the selection of roofing materials are:
— the pitch or slope of the roof
— the desired appearance
— fire resistive aspects
— desired longevity
— cost

ROOF PITCH OR SLOPE:

As stated previously in CHAPTER VIII, **all** roofs must have some slope (pitch) in order to cause water to run off and avoid *ponding*. Different roofing materials however, have differing requirements as to the allowable (max. or min.) amount of roof pitch. For example, most shingle types of roofing (asphalt, fiberglass, wood, slate, etc.) cannot be used on roofs which have very low slopes or little pitch, because they are materials which overlay each other with no true waterproof seal of the joints between them, thus requiring enough slope for water to quickly drain off, as well as to prevent wind-blown water from backing uphill under the layers. Conversely, most asphalt or coal tar pitch built-up *hot* roofs should not be used on steeply sloped roofs, because these systems tend to flow or creep when the asphalt becomes heated by the sun, and can actually slide off the deck, or gather into folds and ridges, tearing and exposing portions of the deck. See (Figure 51) for general guidelines as to the allowable slopes of asphalt roofing products.

APPEARANCE of the roofing materials is a concern for roofs which have sufficient slope to render the finished roofing visible

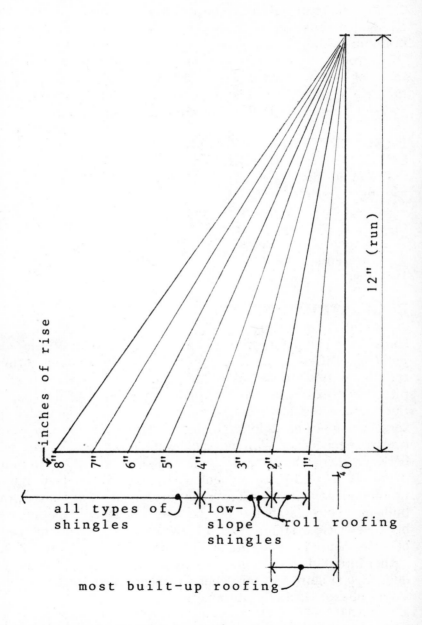

NORMAL ROOF SLOPES FOR ROOFING PRODUCTS

FIGURE 51

to normal view. The general geometry—or shape—of the roof is an earlier decision related to the architectural character or design of the building as a whole. See CHAPTER VIII, and (Figure 45).

Color, texture and pattern of the materials are the three major appearance variables. Wood *shakes* or shingles, for example, will provide a very rustic and dramatic texture, but with limited color range; whereas, asphalt or fiberglass shingles can produce a wide range of textures depending on shingle weight and profile, because they come in varying patterns or cuts, as well as a wide range of surface colors. Tiles of clay, cement and slate all have a unique character, pattern and texture which may dictate or preclude their selection. Metal roofing can be quite plain, such as the often-seen corrugated metal, or be very unique due to a particular pattern of seaming or jointwork between adjacent panels. Metals can also be quite variable in color—either as a natural material such as copper or aluminum, or via coatings, paints, and other processed finishes.

FIRE RESISTIVITY:

Roofing materials may or may not be required by local code to possess some degree of resistance to the effects of fire, relating to their combustibility as well as their resistance to spread fire via flying brands (the release of burning splinters or pieces). Wood shakes and shingles are extremely susceptible to burning. These materials can become the initial source of fire for a building when flying brands from an adjacent field fire, forest fire, or building fire land on the shingles, igniting them. This is an all-too-common problem in many areas of southern California where forest and brush fires spread to residences via their exposed wood shake roofs. We have all seen the news stories showing homeowners and firefighters desperately hosing water onto these kinds of roofs, hoping to prevent disaster. Some building codes will require special asbestos underlayment felts and/or special roof deck construction under wood shakes and shingles, to partially compensate for the high fire risk aspects of this roofing.

Asphalt and coal-tar pitch are by themselves also quite combustible materials; however, when incorporated into the various composite shingle, roll-roofing and built-up roofing assemblies with mineral or stone top surfaces, these assemblies can achieve a greater degree of fire resistivity than wood. They are therefore more readily acceptable by codes for most residential construction applications. Asphalt shingles are usually classified by Underwriters Laboratories (U.L.) as Class "C". This classification indicates they are *effective against light fire exposure*, that is they are *not readily flammable and do not readily carry or communicate fire; afford at least a slight degree of heat insulation to the roof deck; do not slip from position; possess no flying brand hazard; and, may require occasional repairs or renewals in order to maintain their fire-resisting properties.*

The newer—and more common—fiberglass roof shingles usually carry a U.L. Class "A" classification, which signifies substantial improvement in the similar aspects of flame spread, flame communication, etc. described above for Class "C", due to the use of fiberglass fiber mats in the base stock of the shingle.

LONGEVITY:

There is an **extremely wide** variance in the anticipated useful life of various roofs, dependent on the following variables:
1) The degree of exposure to sun, rain, cold and temperature fluctuations.
2) Make-up of the system as to the number of plies or layers, thickness, and weight.
3) Whether the materials are organic, such as wood, asphalt, etc. or in-organic such as metal, clay, slate, etc.

The following table is included with the intent of providing **only a very general** indication, strictly in the opinion of the Author, of anticipated useful lifes of various roofings:

Roofing System or Material	Hot, Sunny Climates	Rainy and/or Cold Climates
• Roll Roofing	6 to 10 yrs.	10 to 12 yrs.
• Light weight Asphalt shingles	10 - 15 "	13 - 18 "
• Heavy Weight Asphalt shingles	15 - 20 "	20 - 30 "
• Built-Up Roofing, (10 yr. Spec.)	6 - 10 "	8 - 10 "
• Built-Up Roofing, (20 yr. Spec.)	12 - 20 "	18 - 20 "
• Wood shingles	25 - 40 "	20 - 30 "
• Slate	75 - 100+"	75 - 100+ "
• Clay or Cement tiles	100+ "	75 - 100+ "
• Non-Ferrous Metals	75 - 100+ "	75 - 100+ "

ROOF DECK:

The base surface, or *deck*, over which the roofing is applied is critical to the success of the roofing. Nearly all residential roofing systems require a deck which is smooth and nailable; that is, that has the ability to be nailed into, with substantial resistance to nail withdrawal. The deck must be strong enough to support the roofing materials plus any other applied loads such as snow, ice, occasional foot traffic, etc. It must not deflect sufficiently under load to cause strain, displacement nor rupture of the roofing materials. Decks usually installed in residential work are plywood, wafer-board, 3/4" tongue-and-grooved wood sheathing boards, or 2", 3", OR 4" thick tongue-and-grooved wood planking. Board and plank decks **must be dry** to prevent warp and shrinkage, which can distort and tear roofing.

For the installation of wood shingle or shake roofing, the deck may consist of 1" x 4" wood boards installed with approx. 4" spaces between each board, in order to allow air exposure to the undersides of the shingles for moisture control and for shingle dry-out. See (Figure 52).

Plywood for decking must be exterior type, or be made with water-resistant glue. The usual grade used is CDX. Required thickness is a function of Code requirements and of the span distance between rafters or supporters, but should not be less than 1/2" for adequate nail holding power.

Any holes, loose knots or other major defects in the decking should be covered with patches, 26 gauge minimum. Iron or steel should be galvanized.

(**CAUTION**) All decks must be clean and free of debris, loose or protruding nails, oil, grease or other contaminants, and be **dry** when the roofing is installed. Moisture which is trapped in materials during construction can be very troublesome and detrimental to the success of any roofing system.

TYPES OF ROOFING, AND INSTALLATION REQUIREMENTS:

A. ROLL ROOFING consists of 36" wide rolls of asphalt impregnated felt which is coated with weather-resistant asphalt, then top surfaced with colored mineral granules. When installed, it forms either a single layer (ply) weighing 80 to 90 lbs. per 100 square feet (a *square*), or 2 plies weighing 110 to 120 lbs. per

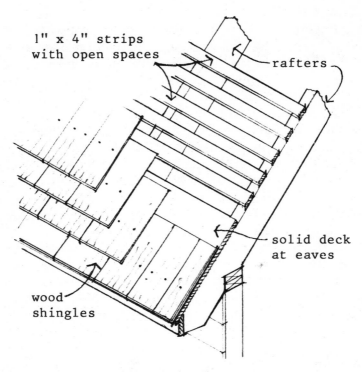

WOOD SHINGLE ROOFING

FIGURE 52

square. The minimum pitch for roll roofing is 1" on 12". The normal range of pitch is from 1" to 4" on 12". Laps are nailed and cemented with special quick-setting cement. Roofing sheets should not be installed when the air temperature is below 45 deg.F.; and, the roofing should be first rolled out, allowed to flatten and be job *climatized* before being installed. All eaves and rakes should be fully cemented to 9" wide continuous strips of the same roofing which have been previously nailed along the edges of the roof deck. Upper edges of roll roofing must be nailed, with the next succeeding course lapping over the nailed portion and being fully cemented to it. All ends of sheets should be lapped at least 6", nailed and cemented. See (Figure 53)

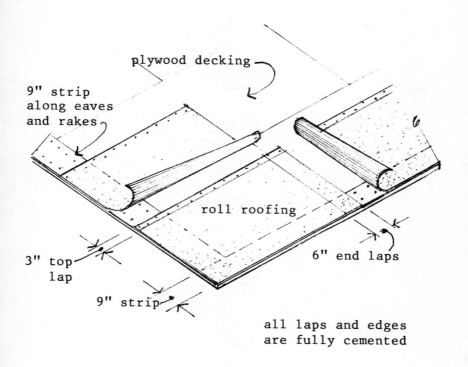

ROLL ROOFING

FIGURE 53

Roll roofing is generally considered to be more of an economical utilitarian roofing system. It does not result in a particularly aesthetic roofing pattern nor texture.

B. ASPHALT SHINGLES are of similar materials to roll roofing; FIBERGLASS SHINGLES differ in that they have a fiber glass mat core. Both are made into separate *shingles*, usually 36″ long x 12″ wide. The exposed edge of each shingle has slits to divide it into sections called *tabs*, which are made in varying sizes and profiles to create different patterns once installed. See (Figure 54) for examples. Shingles come in varying weights and thicknesses, from approx. 235 lbs. to 300+ lbs. per square. Shingles which are *self-sealing* should be used. These have small patches of mastic placed by the manufacturer on the underside of the tabs of each shingle, which when exposed to the warming action of the sun, adheres the tabs to the shingles below, thereby

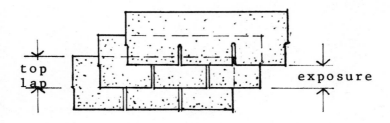

3-TAB, SQUARE BUTT SHINGLES

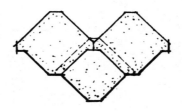

INDIVIDUAL HEX SHINGLES

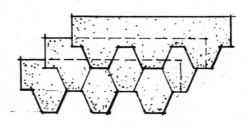

3-TAB HEX STRIP SHINGLES

FIGURE 54

preventing lifting and possible tearing of tabs due to wind. Heavier weight shingles are generally more durable and resistant to wind damage, dry-out, cupping, etc., than the light weights.

(**CAUTION**) Shingles should be installed at roof pitches greater than 4″ per foot. However, there are some shingles made for low slopes between 2″ and 4″ per foot, provided they are installed in accordance with special *low-slope* specifications.

An underlayment layer of No.15 asphalt saturated felt must be nailed to the entire roof deck, before application of shingles.

C. BUILT-UP ROOFING consists of layers of asphalt saturated fiberglass or asbestos felts alternated between field applied coatings of heated asphalt. See (Figure 55).

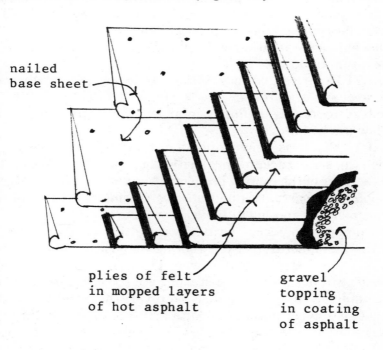

4-PLY, GRAVEL SURFACE, BUILT-UP ROOFING ON NAILABLE DECK

FIGURE 55

From 3 to 5 plies of felt are installed depending on the desired life of the roof, and its particular application. Roof pitch must be an absolute minimum of ¼" per foot, and can go up to 6" per foot, depending on the surface treatment and the type of asphalt used. Pitches of ¼" up to 2" per foot are more commonly found. The steeper pitches require use of *steep pitch* asphalt, which is of stiffer consistency and higher softening temperatures so as to resist the tendency to soften and flow as it gets heated up by the sun.

Built-up roofing receives either gravel top surfacing, a mineral granule surfaced top sheet, or special asphalt top coatings left *smooth surfaced*. Smooth surfaced roofings should have aluminum or white paint-type reflective coatings applied to them. All of these surfaces are intened to prevent deterioration of the roofing by the sun and weather elements, and to reduce surface temperature build-up.

(**CAUTION**) It can be very difficult to locate and repair leaks in a **gravel surfaced** roof.

Built-up roofing is definitely intended, and more suitable, for very low-pitched roofs.

(**CAUTION**) It is very important that built-up roofing be applied in strict accordance with the specifications of the manufacturer of the roofing products. Any variance from this can void potential warranties and absolve the manufacturer of any responsibility for roofing difficulties or failures.

It is possible to secure written warranties, or a surety bond from the manufacturer of the roofing materials, but these will result in increased costs to the owner. It is also possible to require a warranty from the roofing installer for a specific period of time (I recommend 10 years min.); however, the validity of this warranty is subject to the good faith and continuance in business of the roofer. It does, however, tend to increase the roofers' care and attention to installation procedures and workmanship.

On wood board or plank decks, a separate layer of building paper should always be nailed down first, before the built-up roofing. This is not usually necessary on plywood decks. The first ply (or layer) of built-up roofing on plywood, should be nailed to the plywood, or thru the building paper to wood board decks. This is to allow differential movement due to the unavoidable swelling and shrinkage of the wood to take place as it absorbs and releases moisture, without the stresses and other effects being transmitted to the roofing membrane because of a too-rigid connection of the two materials. See (Figure 55).

D. CLAY TILE, CEMENT TILES, AND SLATE: These are all much heavier, durable and more expensive roofing materials. The roof structure may need additional strength to support these materials. All should have at least a 30 lb. roofing felt underlayment, doubled at ridges, hips, valleys and other break points in the roof. All should have nail holes already provided in the tiles by the manufacturer—**not** be punched in at the jobsite.

All nails and fasteners for these materials should be non-ferrous copper, aluminum or be of heavily zinc coated steel.

Some highly-profiled clay tile roofs, like *Mission* and *Roman* tiles, require horiozontal or up-and-down strips of wood applied to the wood deck, in order to properly support and maintain the design profile of the tiles. See (Figure 56).

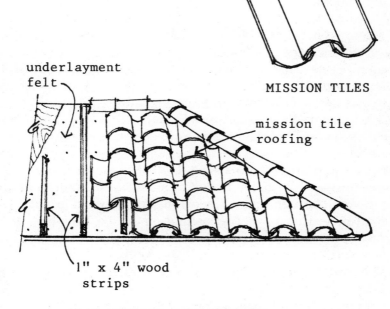

MISSION TILES

mission tile roofing

underlayment felt

1" x 4" wood strips

CLAY MISSION TILE ROOFING

FIGURE 56

(**CAUTION**) Tile and slate roofings are brittle and therefore very susceptible to damage from physical abuse, especially foot traffic on the roof. If there is need for regular access to the roof to perform routine maintenance on equipment, or similar reasons, other kinds of materials should be used in those traffic areas, or other precautions taken to preclude damage to the tile roof.

FLASHINGS:

Flashing are the materials and methods used to make watertight: 1) all penetrations thru the roofing, 2) any intersections of the roofing with other materials, and 3) any intersecting planes of the same roofing.

It is beyond the scope of this book to discuss the nitty-gritty details of various flashing conditions. However, we will illustrate some typical flashing situations and discuss certain principles.

Flashing materials include thin gauge metals (copper, aluminum, lead, tin, stainless steel, and galvanized iron or steel), roofing felts, rubber, fiberglass and various types of asphalt and roofing mastic.

All metal flashings expand and contract due to temperature changes at significantly higher rates than most other roofing materials they are used with. Therefore, provisions must be made to allow metal flashing parts to move. If they are confined or constrained by rigid fastenings or joints on all edges—especially in long lengths—the internal stresses will cause metals to fold, crease, ridge and eventually crack open.

The basic principle of flashings is that parts at higher elevations must lap over the top of parts at lower elevations—not the other way around. This is so as to shed water away from the joint, not into and under it. Joints must be so made that water could work through them only against the force of gravity.

Typical locations where flashings must be provided are the following:

— Penetration of a chimney thru the roofing (Figure 57),
— Penetration of plumbing vents and other pipes thru roofing (Figure 57),
— Valleys in roofs (Figure 58)
— Roof ridges and hips (Figure 59)
— The intersection of roof with vertical wall or parapets (Figure 47)

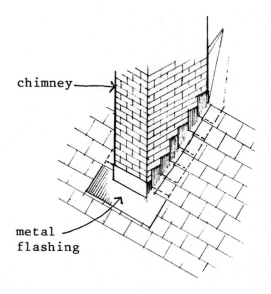

CHIMNEY FLASHING

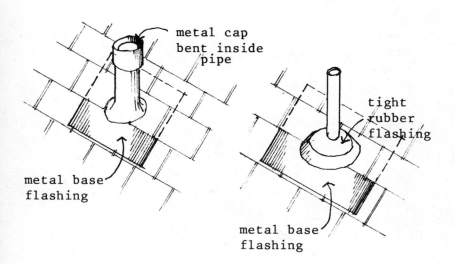

FLASHING OF VENTS AND PIPES

FIGURE 57

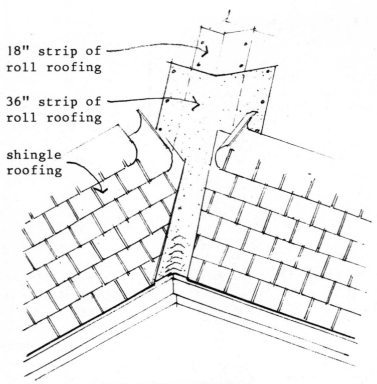

OPEN VALLEY FLASHING

FIGURE 58

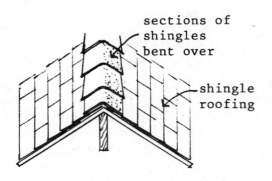

RIDGE FLASHING
(HIP SIMILAR)

FIGURE 59

ICE DAM FLASHING:

In cold climates, buildings with roof overhangs can experience troublesome leaks due to a phenomenon known as *ice dams*. In these instances, snow and ice which collects on the roof will gradually melt from the heat which escapes from the building interior. The melted water runs down towards the lower edges of the eaves. If the outside air is still quite cold, it is likely that the water will turn to ice again on roof overhangs which extend beyond the outside walls, because there is no interior heat source at these locations. This ice builds up into an *ice dam*, causing water to back up the roof, into and under the roofing materials, then into the unprotected building structure. Significant damage has been done to many cold-country buildings because of this phenomenon. There are two generally used alternative methods of eliminating this problem. One is to use watertight metal roofing on the overhangs, extending the metal back up the roof to at least 12" inside the interior wall line, where it is flashed to— and covered by—the regular roofing material, such as shingles, tile slate, etc. The other alternative is to install a 2-ply underlayment of No.15 asphalt saturated felts and asphalt cement from the drip edge of the eave back up to at least 12" inside the interior wall line for roof slopes over 4" on 12", and at least 24" inside the interior wall line for roof slopes less than 4" on 12". The finished roofing material is then installed in the regular manner over the entire roof, including the special 2 ply underlayment. See (Figure 60) for examples of both alternatives.

GUTTERS AND LEADERS

Gutters, sometimes called eavetroughs, are installed at the eave drip edges of roofs, to catch run-off water and carry it to leaders, or downspouts, which deliver it to the ground at desired locations. Gutters thus prevent random, uncontrolled dripping of water from roofs, which causes soil erosion as it splashes on the ground, as well as discoloration and deterioration of the lower portions of the building.

Gutters are made of metal or wood, and may be attached to the fascia at the outside edge of the eave, or be integrally built into the eave construction, otherwise known as a "concealed gutter". See (Figure 61).

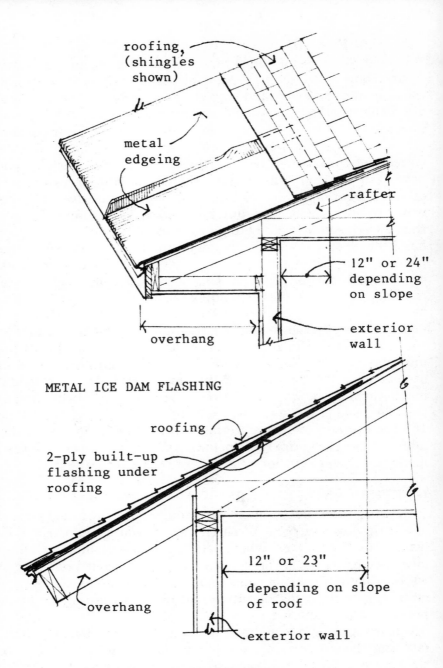

FIGURE 60

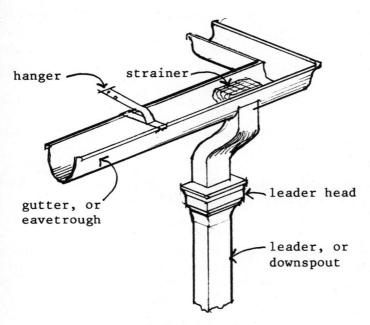

ATTACHED GUTTER

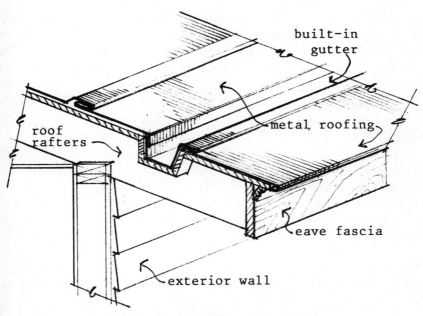

CONCEALED, OR BUILT-IN GUTTER

FIGURE 61

(**CAUTION**) Built-in gutters must be very carefully lined, flashed and sealed with permanent non-ferrous materials, or else they will eventually develop leaks and cause serious deterioration to the overhang structure, exterior walls and finishes.

Gutters must be:
1) Large enough to handle the quantity of water being discharged.
2) Sufficiently pitched to carry the water off quickly, and not retain pockets of water.
3) Leak-proof.
4) Installed so as to avoid possibility of snow and ice backing up under the main roofing material.

Leaders are vertical pipes which carry the water from the gutter to the ground, or to a ground level drain. They are usually made of metal, in various shapes; but, they must be sized to carry the water away as fast as it is received, so as not to cause gutter over-flow.

WIND FORCES AND UPLIFT ON ROOFS

Wind striking a building is deflected upward, passes over and around the building. In so doing, wind produces areas of reduced pressure over the surfaces of flat or slightly inclined roofs. This pressure can be considerably lower than the pressure inside the building. This differential between inside and outside pressures tends to lift the roof. Uplift force from suction is highest along roof eaves and corners which face the wind, for flat or slightly inclined roofs. It is also high on the leeward slope of steeply inclined roofs. See (Figure 62).

(**CAUTION**) As a result of these forces, roof structures, decks and roofing materials which are not securely fastened are subject to lifting and possible blow-off. Special care must be taken along eaves and edges to see that a sufficient number of fasteners have been used to secure the edging to the building, since this a most vulnerable area. Similar precautions apply at the corners of roofs.

(**CAUTION**) The fastening recommendations of roofing manufacturers **must be followed**, and should be considered as minimum requirements for the attachment of roofing materials. Their specifications and recommendations are based on many years of field experience as well as testing. Most manufacturers print these instructions on the wrappings used to package their materials. They are also readily available from any major roofing supply house. The Author has personally witnessed major blow-off of built-up and shingle type roofing installations, which when analyzed were determined to be a direct result of a lack of adherence to the fastening requirements of the manufacturer. It is much better to error on the side of fastening over-kill on this issue.

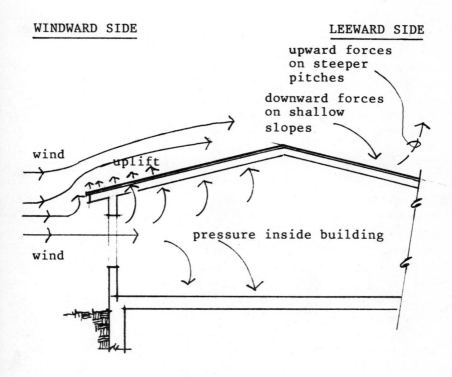

DIAGRAM OF EFFECTS OF WIND FORCES ON ROOFS

FIGURE 62

CHAPTER X... INTERIOR CONSTRUCTION AND FINISHES

PARTITIONS:

A partition is an interior wall which divides, sub-divides or encloses space. Most partitions extend full-height from floor construction to overhead ceiling or roof construction, in order to create completely separate rooms which allow different functions with privacy. However, partitions can be less than full-height; they may not fully surround new rooms or spaces; they may contain glass panels; etc.

Partitions may be constructed of many different materials, including wood stud framing, plywood panels, masonry, bookcase type millwork, etc. Most common is the 2" x 4" wood stud framed partition, built almost exactly like the previously discussed wood framed exterior walls, with a base shoe or sole plate at floor line, vertical studs, and a cap plate at the top.

Interior partitions may be bearing or non-bearing. If bearing—i.e., supporting overhead floor and/or roof construction—they are usually located directly over supportive construction below, such as a beam, girder, or another load-bearing wall. The top plate or cap on load bearing partitions should be a double member, such as two 2" x 4"s, to better support the loads. See (Figure 32).

(**CAUTION**) Oftentimes, attempts will be made to save floor space by constructing thinner partition(s) with the studs turned the 2" way. This practice results in a *flimsy* wall which does not have adequate lateral strength for its un-supported height; the only locations where this should be accepted would be as a divider between adjacent closets, furring around a pipe space, or other similar, very limited, purposes. Basic room partitions and any partition with a door in it should never be of *thin* stud construction. See (Figure 63).

In order to enclose plumbing pipes, ducts or other mechanical devices, partitions occasionally must be constructed of 2" x 6" or 2" x 8" studs, or be two slightly separated stud partitions. See (Figure 64).

All door openings should have doubled studs at both sides (jambs) and top (head). Stud *rough* openings are made approx. 2" larger than the door size, to allow for the door frame thickness

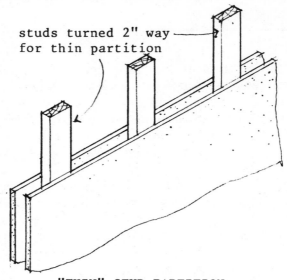

"THIN" STUD PARTITION

FIGURE 63

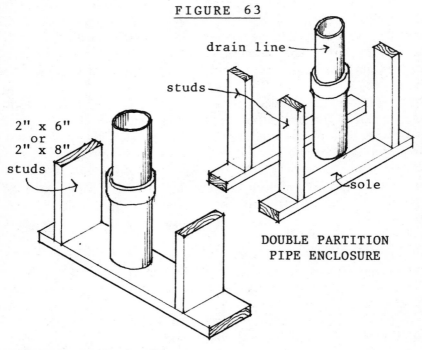

WIDE STUD PARTITION
PIPE ENCLOSURE

DOUBLE PARTITION
PIPE ENCLOSURE

FIGURE 64

plus provide clearances to shim, adjust and plumb the finished frame. See (Figure 65).

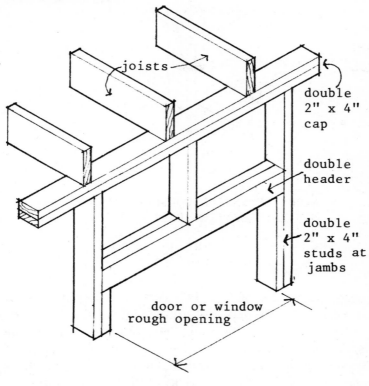

FIGURE 65

Partitions should always be constructed so that all edges, ends, tops and bottoms have a solid nailing member. This requires additional studs and framing at interior and exterior corners and at intersections with other walls. See (Figure 66).

INTERIOR STAIRS:

As pointed out in CHAPTER V... FLOOR SYSTEMS, openings in wood framed floors for stairs or other purposes must have double joists on all sides of the opening, to provide the proper floor strength and stiffness at the opening. See (Figure 33).

If stairs are constructed in place at the jobsite, they should have rough *carriages*, or *stringers*, which are made of 2" x lumber cut to the stair profile. There should be three (3)

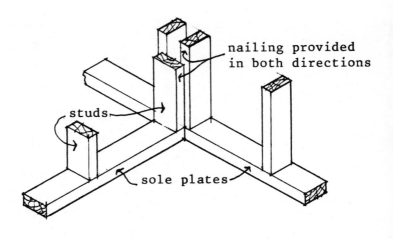

**EXAMPLE OF INSIDE CORNER FRAMING
AT INTERSECTING PARTITIONS**

FIGURE 66

carriages for stairs 32" to 44" wide; two (2) may suffice for narrower stairs. The depth of lumber left in the carriages below the cut-out for the steps must be adequate to carry the stair loads. See (Figure 67).

Finished stairs may be quite plain and simple, or be very ornate and decorative. Stairs with exposed sides, highly decorative parts, or complex construction will almost always be built off the jobsite by a special millwork firm, and be brought to the jobsite ready to set in place. See (Figure 68) for terminology of the various parts of a stair.

The intricate details of interior stair construction are beyond the scope and purposes of this book. However, following are some guidelines for proper stair proportions and function.
1) There should not more than 1/8th of an inch variation in the height of any single stair riser, or the width of any single tread; nor, more than 1/4" variation in height between all risers.
2) The relationship of riser height to length of run of each tread must comply with local Codes. Many formulas and criteria are in use to assure good stair proportions for safe, comfortable use. Some examples are:

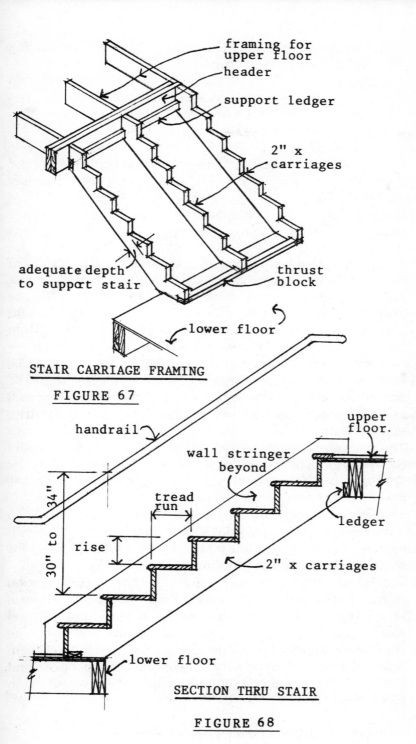

STAIR CARRIAGE FRAMING

FIGURE 67

SECTION THRU STAIR

FIGURE 68

- Maximum riser heights: varies from 7½" to 8¼", depending on which code or guide is consulted.
- Minimum Tread Width (Not including the nosing): 9" to 10", depending on which code or guide is consulted.
- 2 Risers + 1 Tread = not less than 24", nor more than 25".
- Riser + Tread = 17 to 17½
- Riser x Tread = 70 to 75.

It is the Author's opinion that the most comfortable interior stairs have risers between 6⅝" to 7⅝", and comply with the Riser x Tread = 70 to 75.

3) Stairs should have a minimum of three (3) risers. Fewer than 3 become stumbling hazards, and should be ramps instead.
4) All stairs should have handrails, preferably on both sides of the stair. Handrails should be located not less than 30 inches, nor more than 34 inches above the tread nosing. See (Figure 68). Handrails must be continuous. One handrail should extend at least 6" horizontally beyond the top and bottom risers. Ends of handrails should bend and return on themselves, or into a wall, or terminate in newel posts. The handgrip portion of the handrail should not be less than 1¼" nor more than 2" in cross-section. Handrails alongside a wall must have a space not less than 1½" between the handrail and the wall.
5) The maximum height of rise for any single flight of stairs between floors or intermediate landings should be 12 ft.
6) Winding and spiral stairs, while allowable by some Codes, are dangerous and should be avoided wherever possible.
7) The vertical headroom between nosings of stair treads to any overhead construction should not be less than 6ft. 6inches; however, this may be visually uncomfortable for some people, and so the Author recommends 7'-0" min.
8) Special care and workmanship is required in the construction of stair parts and joints, to create stairs that will not loosen up over time and *squeek* or become noisy due to poor fits and movement. This usually means that treads and risers be connected together by tongue and grooves, and both of them be housed into cut-outs in the side of stringers, all nailed and glued together and made tight by wedges nailed and glued in place. See (Figure 69).

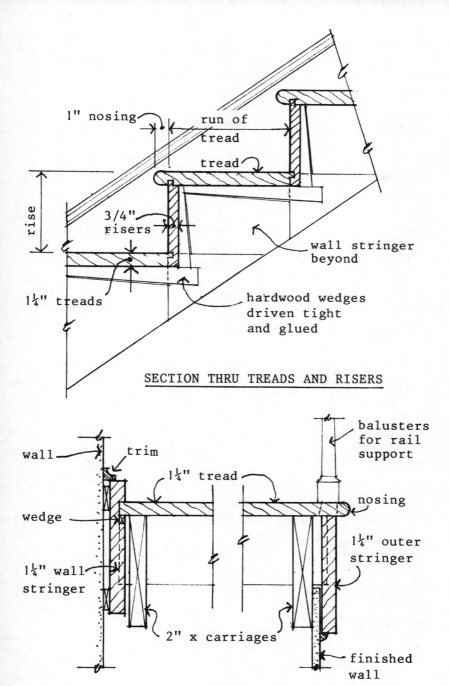

FIGURE 69

LATH AND PLASTER FINISH:

Prior to the widespread use of gypsum board *drywall* finishes which began in the late 1940's and early 1950's, lath and wet plaster were the principal interior wall and ceiling finish materials in use. There is still some wet plaster installed today, indeed, some people believe it results in a more durable, tough and desirable finish. The chief drawbacks are the somewhat higher cost, the less availability of skilled craftsmen, and the fact that it is wet construction which creates high moisture levels in the building during the dry-out period, which can be a problem because of its negative impacts on other finishing operations, on millwork, doors, cabinetry, etc. Plaster is also more prone to cracks due to shrinkage, etc., than gypsum drywall.

LATH is the foundation for plaster. It must provide a good base to which the plaster will permanently bond, as well as add strength and crack resistance to the plaster. The common laths in use today are metal, which has been expanded or processed to provide many openings thru which the plaster will *key* or attach itself (See Figure 70);

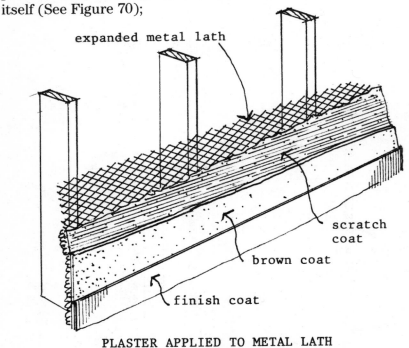

PLASTER APPLIED TO METAL LATH

FIGURE 70

and, gypsum board lath which is similar to gypsum drywall board and may be smooth faced or contain a pattern of holes to improve mechanical bond of the plaster.

(**CAUTION**) Strips of expanded metal lath must be installed to reinforce all corners and intersecting surfaces of gypsum lath, and to improve crack resistance (Figure 71).

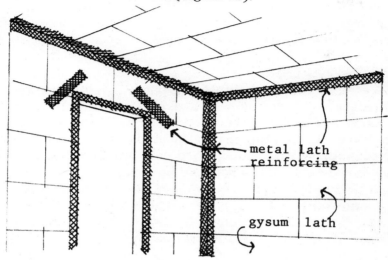

**GYPSUM LATH BASE FOR PLASTER
REINFORCED WITH METAL LATH**

FIGURE 71

PLASTER is a mixture of sand, water and either hydrated lime or gypsum mortar. Gypsum is the more prevalent material in use today because it is less troublesome to mix and to apply.

Plaster is applied to the lath in either 2 or 3 separate layers, or coats. The 3rd, or finish coat, can be made very smooth or be given a variety of different textures.

(**CAUTION**) In kitchens, baths, and other areas where there will be high humidity and dampness, a special type of plaster called *Keenes Cement* should be used. When set, this is very hard and more resistant to the effects of moisture. It is not suitable, however, for shower walls or other areas subject to constant wetting.

Chapter 10: Interior Construction and Finishes

Thin coat plaster consists of approx. 1/8" to 1/4" of finish coat plaster which is applied to gypsum board. It results in a more dense, hard wall surface, thus providing some of the characteristics and benefits of plaster without the higher cost.

Plaster may be successfully applied directly to clean masonry surfaces. Surfaces which are too dense or smooth to allow a good bond, such as smooth formed concrete, should first have special self-furring metal lath mechanically attached thereto, to form a suitable base for the plaster to bond.

DRYWALL FINISHES:

There are many substitutes for plaster, all of which result in a finished wall surface which avoids the wet processes and mixtures. They involve the use of pre-formed rigid board materials applied directly to the studs or framing, and include gypsum drywall, plywood, wood pulp and fiber boards, cement-asbestos boards, and solid wood paneling.

A. GYPSUM BOARD DRYWALL. Called *gyp-board*, this has a gypsum core faced with paper on all faces and edges except the ends. Sheets are commonly 4ft. wide, from 8 to 14 ft. long and from 1/4" to 5/8" in thickness. The long edges are tapered for receipt of joint tape and finishing compound. Gyp-board is also made in ASTM Type 'X' which has a more fire-resistive core material, and in 'WR' water-resistant form, commonly called *greenboard*, (because of its distinctive green paper cover) for use in damp or moist locations. The term *Sheetrock* of often used to refer to gyp-board; however, this is actually the trade name of United States Gypsum Co. for its gyp-board.

Typical residential interior use of gyp-board is a single layer, not less than 1/2" thick, fastened by special dry-wall screws to the wood framing, or nailed with special gyp-board nails. Screw and nail heads must be driven slightly below the surface of the board so they will be concealed when covered with compound. On framing supports up to 24" on centers (o.c.), sheets may be applied with long edges either parallel or perpendicular to framing members. The latter has gained more recent acceptance because it seems to provide a greater strength and compensates somewhat for uneven framing alignment.

If fastened by nails, double nailing is recommended, in which the first nails are spaced at approx. 12" on centers throughout the panels, followed by a second set of nails approx. 2″ from the first. This seems to provide greater holding power and resistance to nail *popping* than normal single nailing. See (Figure 72). Screw fastening provides the most secure attachment, however.

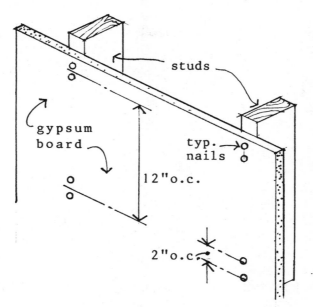

DOUBLE NAILING OF GYPSUM BOARD

FIGURE 72

All joints and inside corners of gyp-board must be reinforced with special reinforcing tape embeded in joint compound, so as to resist cracks and joint show-through (Figure 73). Outside corners should be reinforced with special metal corner bead to resist physical damage (Figure 74).

Joints and nail heads are finished with pre-mixed joint compound made especially for the purpose.

(**CAUTION**) In order to achieve proper coverage, smoothness and hiding power, there should be three (3) separate applications of the compound, with preceeding coats allowed to dry before the next coat. Some fine sanding and smoothing may be required of each coat.

(**CAUTION**) Gypsum board does not afford any significant amount of holding power nor bearing support for fastners.

Chapter 10: Interior Construction and Finishes

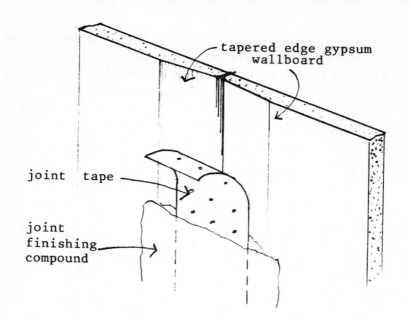

FIGURE 73

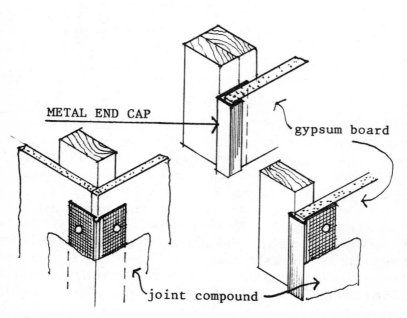

FIGURE 74

Therefore, wood blocking and backing should be installed during the framing process of the building for the future attachment of all items such as cabinets, mirrors, bath accessories, towel bars, tissue dispenser, etc. See (Figure 75). Reliance on *molys, toggle bolts* and such fastners which secure only to the gypsum board, is highly prone to eventual pull-out, working loose, etc., and should be avoided.

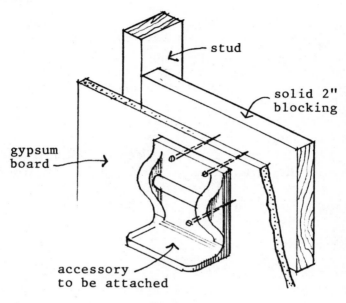

SOLID BLOCKING BEHIND ATTACHED ITEMS

FIGURE 75

Wherever gyp-board terminates or abuts against another material with no subsequent trim or other covering to be installed over the joint, a pre-formed metal or plastic *casing bead* cap should be installed over the end of the gyp-board to make a properly finished, neat termination (Figure 74). Joint compound should not be used to fill and close the gap, as it will eventually shrink away from the other material and become an un-sightly crack.

B. PLYWOOD, WOOD PULP AND OTHER DRYWALL PANELS can be nailed and/or glued directly to wood framing in order to achieve different surface appearances and textures. Many, such as pre-finished plywood panelings and vinyl covered gypsum boards, do not require further finishing operations when installed, except for possible treatment of the joints between panels.

Plywood should be 1/4" thick, minimum, for framing spaced up to 16" o.c. Framing in excess of 16" o.c. requires either increased plywood thickness, or the installation of 16" o.c. cross furring to the studs or joists. All edges and ends of plywood and other drywall panels should be supported by solid framing or blocking.

If the joints between panels are not going to be covered with trim, finish nails should be used, with the heads *set* or driven below the surface. To improve their appearance, such joints may be slightly sanded, shaped into a Vee, covered with strips of matching or contrasting wood, called battens, or treated with a variety of other trim shapes available for the purpose. In these cases, special matching colored nails should be used, or small finishing nails which are set and the holes filled with colored putty or filler. See (Figure 76) for various types of joint treatments.

SOUND TRANSMISSION:

The ordinary stud wall with gyp-board or paneling surfaces does not afford much resistance to the passage of sound thru it. A slight improvement in resistance can be achieved by the installation of fiberglass batt or blanket insulation into the stud cavities. However, a significantly better sound resistive wall is achieved by constructing two separate, **un-connected** partitions parallel to each other, with an air space between. Insulation can be installed in one or both faces, or—better yet—be run continuously between the two faces, being attached to only one.

Care must be used in all instances to be sure that there are no penetrations thru both walls such as doors, ducts, pipes, electrical boxes, etc. If any be required, they should be especially constructed, baffled, gasketed and caulked to be sound transmission resistive. Electrical outlets and switches must not be directly back-to-back in the two faces of partitions which are sound sensitive, and the openings and backsides of any outlets should be

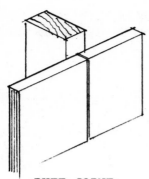
BUTT JOINT

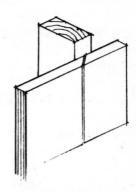
MITERED JOINT

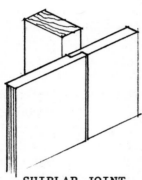
SHIPLAP JOINT

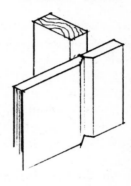
VEE JOINT

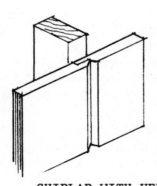
SHIPLAP WITH VEE

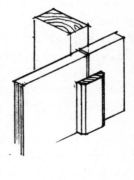
BATTEN

TYPES OF JOINTS BETWEEN PANELS

FIGURE 76

sealed and wrapped with insulation.

DOORS, FRAMES AND HARDWARE have been previously discussed in CHAPTER VII ... WINDOWS, GLASS, DOORS AND HARDWARE.

WOOD TRIM AND MILLWORK:

Millwork comprises all finished woodwork including cabinets, casework, and trim.

Trim includes any exposed finished members such as door casings, baseboard, chair rail, picture mold, ceiling trim or cornices, fireplace mantle and trim, etc. Trim is used to:

1) Cover and finish the joint between surfaces such as the baseboard at the juncture of wall to the floor, and casings over the joint between door frame and the wall;
2) Accentuate certain lines or surfaces; and,
3) Serve utilitarian functions such as a chair rail to prevent abrasion of walls from movement of chairs, or a picture mold from which to hang pictures.

Wood for trim should not have knots, streaks of pitch or resin, and should be fine, close-grained. If the trim is to be painted, Ponderosa Pine, Idaho white pine, yellow poplar or other so-called whitewoods are good choices. These woods will sand to smooth finish and will retain good tight fitting joints, if of dry stock. Longer lengths of various trim members will often consist of short pieces of wood which are *finger-jointed* together, glued and sanded smooth at the mill. These do result in lengths which are quite sound and acceptable as painted trim. Finger-jointed or other processed trim which is made up of shorter sections of different pieces of wood are not recommended for natural or stained finishes, because the different pieces and their joints will be very pronounced, discordant and unsatisfactory in appearance.

Trim which will be natural or stained finished, is usually best made of hardwoods such as oak, birch, maple, walnut, teak, cherry, or mahogany. These woods have very pleasing decorative grains and contribute a sense of high-quality and richness to the interior spaces.

It is important that trim and millwork items be straight, and be kept covered and **dry** when stored at the jobsite. They should not be permitted in the building nor be installed until the interior has dried out from any wet processes such as plastering, or concrete work.

Trim which will be painted should have its primer coat of paint put on at the mill before it is shipped to the jobsite. This assists in the prevention of moisture being absorbed into the wood. All painted trim should be primed on its backsides as well as the exposed finish surfaces, to resist moisture absorption.

In the installation of trim, care exercised in creating joints and tight fits makes substantial differences in the appearance. Mitered joints are almost always preferable to butt joints. Ex: when separate lengths of material must be used to make up a long continuous trim member, the joints between ends should always be mitered, so that if these joints open up slightly due to shrinkage, they do not show as unsightly cracks as do butt end joints. See (Figure 76).

Molding and trim can be purchased in many molding profiles. Various moldings can be combined, as are more frequently used in the *classic, victorian* and *traditional* house styles. See (Figure 77).

WAINSCOTS:

A wainscot is a covering or finish of the lower portion of a wall with a material or materials which are different from the upper wall. Wainscoting may be done strictly for the sake of appearance, or for functional reasons, such as to provide greater resistance to wear and abuse, more moisture resistance, etc.

A wainscot can be of very durable and water resistant material such as ceramic tile, which may be very desirable in baths, kitchens, laundry rooms or spas. It can be of plywood or solid wood paneling, attached to the basic wall surface or to furring members. Such materials usually require separate trim to finish and cap their top and bottom edges. See (Figure 78).

FLOORING:

A. WOOD FLOORING is primarily hardwood (oak, maple, cherry, pecan, teak, birch and beech) for required hardness and durability. It is available primarily in two types:

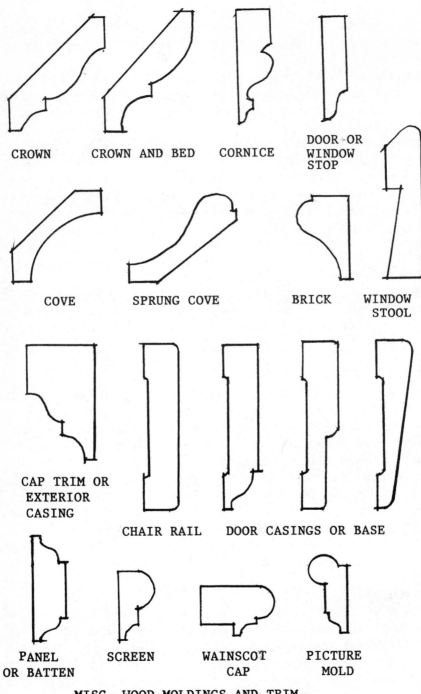

MISC. WOOD MOLDINGS AND TRIM

FIGURE 77

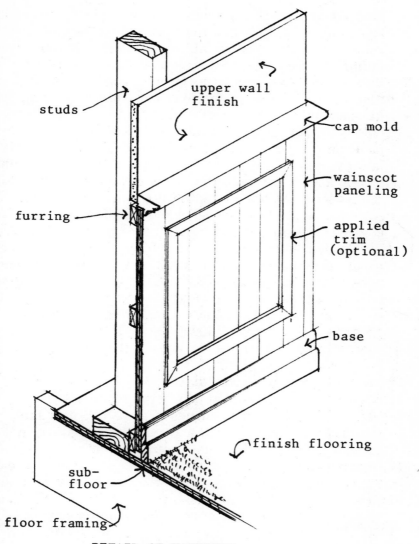

DETAIL OF WAINSCOT

FIGURE 78

1) **Strip**, or **plank**, flooring which is made in face widths of 1½", 2", 2¼", 3", 4", 5", 6", and 7" (the most commonly available are 1½", 2", and 2¼"). Thicknesses are 3/4" or 25/32", and random lengths. All edges and ends should be tongue and

grooved for good fit and alignment between adjacent strips. Fastening is by concealed toe nailing thru the edges to the wood sub-floor. Almost all strip/plank flooring comes unfinished, requiring sanding and finishing after installation.

2) **Parquet** flooring, consisting of strips of hardwood glued together by the manufacturer into blocks having face dimensions of 6"x6", 6"x12", 9"x9", 12"x12" and 16"x16"; however other sizes are available, depending on the manufacturer. The common thicknesses are 5/16" and 3/4". There are a variety of wood patterns available within individual blocks, and overall patterns can be achieved depending on the pattern and orientation of alternate blocks when laid. All edges and ends of each block should be tongue and grooved for good alignment and surface control. Fastening is usually by mastic or glue to the subfloor. Parquet flooring is usually pre-finished at the factory. Some parquet is available with a thin (approx. 1/32" thick) layer of closed-cell polyethylene foam on the backside, to act as a cushion for reduction of impact noise as well as a moisture barrier.

The key factors to look out for when installing strip flooring are:
1) The flooring must be **very dry**;
2) Flooring must be installed with very tight joints;
3) Nailing must be frequent enough (8 to 12" o.c., max.) to force the flooring into tight contact during installation, and to prevent loosening or movement after installed.
4) Most manufacturers recommend a layer of building paper or felt be placed down over the sub-flooring prior to installing strip or plank flooring. This improves moisture resistance, and assists in keeping floors from making squeaky noises.

B. RESILIENT FLOORINGS include sheet goods such as linoleum and sheet vinyl, plus individual square resilient tiles such as vinyl, vinyl asbestos, rubber, and asphalt tile. These synthetic floorings are of thin gauge material (1/16" to 3/16" thick), intended for floors which are durable, inexpensive and readily cleanable from spills and other soiling. The sheet materials come in 6ft. and 9ft. wide rolls, and when installed have the distinct advantage of very few joints. Resilient tiles are available in 9"x9" and 12"x12" sizes. Some tiles tend to shrink slightly over time, making joints more pronounced, thus allowing dirt to collect and water to penetrate.

Resilient floorings are installed by special cements troweled on to the sub-flooring. It is highly recommended that an underlayment such as particle board or special underlayment-grade plywood (3/8" min. in thickness) first be installed over the subfloor, to achieve a perfectly smooth, puncture-resistant base for resilient flooring. Some manufacturers also specify a layer of building paper be installed directly under the resilient flooring; check the printed recommendations of the manufacturer on this. Some—but not all—resilient materials can be cemented directly to concrete floors; this varies depending on whether the floor is below grade, at grade or above grade.

C. CARPET can be installed directly over wood or concrete sub-floors. Non-cemented installation, over a good quality pad is recommended. In this, the carpet is attached and held in place via special wood *tack strips* which are fastened to the sub-floor all around the room perimeter. The carpet is stretched and hooked onto the tack stripping, with the edges of the carpet then tucked under. See (Figure 79).

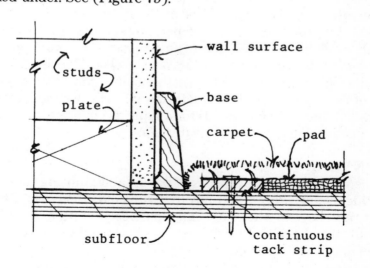

CARPET AND PAD, TACK STRIP INSTALLATION

FIGURE 79

D. CERAMIC TILE for floors should be non-glazed tile with slip-resistant surface texture and finish. Tile for walls and other surfaces can be glazed or non-glazed. Many tile sizes, shapes, patterns and colors are available. Most ceramic tile is approx. 1/4" thick; smaller face sizes usually come pre-mounted on flexible backing material, arranged with the proper spaces between tile for grouting. The tiles are installed with the backing material left in place. Larger tile sizes are usually in individual pieces.

On concrete slab floors, the most common installation is via water-resistant thin-set adhesive. Joints between tiles are later filled with water-proof grout, and then sealed.

On wood framed floors, tile can be cemented directly to an underlayment layer, similar to resilient tile; however, this is not a good method where substantial wetting of the floor is likely, since major water exposure will eventually cause deterioration of the underlayment and sub-flooring. On wood framed floors ceramic tile which requires an ability to withstand significant wetting should be placed on a concrete or cement-mortar setting bed which is built upon the wood sub-floor. A layer of waterproof material is required under the setting bed to keep moisture from penetrating to the wood. This installation can result in being from 1 3/4" thick, or greater, depending on floor variations, how much pitch is provided for drainage, etc. This can present a problem in meeting and matching the levels of flooring materials in adjacent spaces, unless the setting bed is depressed below the top of floor joists, or the floor framing in the area is lowered or made thinner. See (Figure 80) for a typical cement-mortar setting bed installation over a wood sub-floor.

PAINTING:

The many technical aspects of paints and painting procedures are highly specialized and involved subjects which, for the most part, are beyond the scope and purposes of this book. Much technical and product advice can be obtained directly from paint specialty outlets or from painting sub-contractors. Herein we will instead attempt to provide only certain significant pointers, such as:
  1) The quality of results achieved in the final appearance of painted, stained or natural finishes is a direct function of how much preparation such as sanding, filling of holes, etc. is

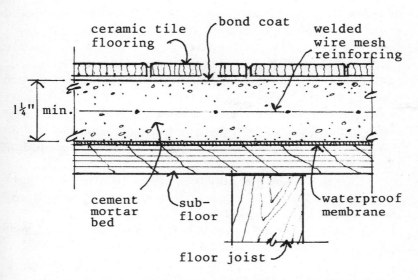

CERAMIC TILE FLOOR ON MORTAR SETTING BED
(FOR WET AREAS)

FIGURE 80

done.
2) 3 coats of paint are better than 2; 2 is better than one; One coat is **never** sufficient to hide or cover properly.
3) Knots and resin streaks in wood must first be sealed with a coat of shellac if oil-based paints are to be used. There may be some bleed-thru with water based paints also, which will require additional coat(s) to hide.
4) Surfaces must be **clean** and **dry** before applying painters finishes.
5) Water-based acrylic or latex paints seem to offer superior resistance to blistering and peeling on exterior surfaces—especially wood based sidings—than the oil or alkyd base paints.
6) Flat latex or acrylic interior finishes are the most commonly used on residential interior surfaces such as walls and ceilings; however, they generally do not stand up well when washed. Many can be actually rubbed off or be badly streaked when subject to brisk washing for removal of stains. Under those conditions, the latex semi-gloss and gloss

enamels are much better than the latex flats. Oil based enamels are the most durable, and easy to clean.
7) Until they dry, water based latex and acrylic paints are much more convenient to clean up from, than are oil based.
8) Very dense, or high gloss, finishes require sanding between coats to improve adhesion and bond.
9) Most manufacturers make more than one grade of paint. It is always better to use *premium* or *specification* grades than less costly grades, because of better coverage, better hiding power, etc.
10) For stained or natural finishes, Polyurethanes are much more resistant to damage from fluids and physical abuse than normal interior varnishes or laquers.
11) Epoxy-based materials are best for paint finishes on very severe-use outdoor conditions, or exposures.

CHAPTER XI... CABINET WORK AND OTHER BUILT-INS

CABINETS:

Cabinets and cabinet-work has today become very much a product of specialty manufacturers. These firms produce many designs and sizes using factory production-line techniques, which are shipped to the jobsite ready to install. It is, then, very rare that cabinets are custom-built at the jobsite, as was the practice several years ago. However, as a result of manufacturing constraints, cabinets are usually only available in certain stock widths, heights, and pre-determined functions (i.e., sink fronts, oven or range cabinets, food pantries, units with all drawers, units with both drawers and shelving, etc.).

There is a wide variance in the quality and cost of cabinets, depending on: 1) Whether they are of wood or metal construction; 2) The quality and species of solid woods used; 3) the type and quality of hardware such as pulls, hinges, locks, drawer slides, etc.; 4) How elaborate the detailing of their face materials.

Almost universally, cabinets today extensively use wood particle board instead of solid woods or plywoods for broad flat surfaces such as sides, ends, tops, and frequently—the shelves. Particle board is a processed material made of wood chips and resins which results in tough, durable and dimensionally stable flat surfaces. It does not have a grain or pattern like natural wood, and thus does not present a very attractive finished appearance by itself. Also, the edges and ends do not cut with a smooth desirable surface; so, the manufacturers commonly cover all exposed surfaces with thin veneers of natural woods or with plastic laminates, either colored or in artificial wood grains. The degree of the laminating process and the relative quality and appearance of the artificial effects is a factor in the cost of the cabinets. There are definite advantages to this product; i.e., it provides surfaces which are truly resistive to the effects of water, grease, detergents and many other household products, plus it results in surfaces which are quite dense and not easily subject to dents or deep scratches. This construction is generally less costly than solid wood or plywood construction. Filler pieces of matching finishes are made in various sizes, to close up left-over openings and provide often needed transitions between different cabinet conditions such as at inside corners, ends, etc. They are job scribed and cut to exact fits.

COUNTER TOPS are made of many finish materials. If of plastic laminate or other thin *sheet* goods, or perhaps ceramic tile, a sub-base is required of plywood or particle board, at least 3/4" thick for rigidity. The edges and ends may be built-up to greater thickness for reasons of appearance and added stiffness. Tops can also be manufactured of homogeneous processed resins which are often used in bathrooms for vanity and lavatory tops. In these, sinks or lavatory bowls can be integrally formed into the top, resulting in a seamless, truly watertight unit. Most resin tops have the appearance of marble.

It is important to select cabinets based not only on appearance or attractiveness, but to look further at detailing and craftsmanship. Operate drawers and doors to observe the quality and *heft* of the hardware. Check to see if drawer fronts and sides are joined by *dovetails* (best), securely screwed (next best), or simply nailed together. Are the drawer sides particle board (poor), or solid wood (best). Do drawer bottoms seem flimsy, or rigid? Shelves for storage should be easily adjustable, in multiple spacings. Do drawers have two side-mounted heavy-duty mechanical slides (best), a bottom-mounted single track slide (next best), a top mounted track guide (next), or simply rest and slide on the edges of the cabinet frame (poorest)? Are the drawers easily removable for cleaning? Are the doors held in closed position by very positive hinge action or magnetic catches, or by much cheaper metal or plastic friction latches?

Other built-in or cabinet items such as bookshelves, bar, special wardrobe closets, laundry work and storage areas, etc. are made of materials and finishes similar to kitchen and bath cabinetwork described above.

CHAPTER XII... CHIMNEYS AND FIREPLACES

CHIMNEYS:

A chimney is an open vertical shaft which carries smoke and hot gasses from a fireplace or furnace up to outdoor air. A chimney operates on the principle that hot gasses rise, being displaced by heavier cold air at their bottom inlet, thus creating a natural upward convection current. A chimney must have some vertical height to create sufficient pressure differential between the incoming colder air at the base inlet of the furnace or fireplace, and the discharge point of the hot gasses, in order to set the vertical convection process in motion. By codes, they also must rise a certain amount above adjacent roof levels to minimize fire hazards. See (Figure 81)

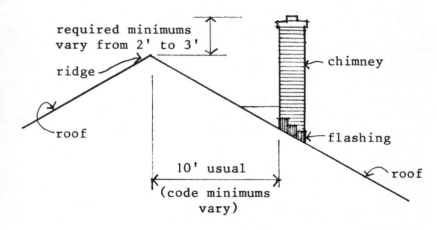

FIGURE 81

Since the hot gasses they contain can be several hundred degrees F. in temperature, chimneys must be built of non-combustible materials. The most common materials are masonry (brick, block, stone etc.) and pre-formed metal.

The chamber which contains the actual hot gasses is called the *flue*. In a masonry chimney, the flue should always be built of "terra cotta" flue lining material, which is dense fired clay made into round, square or rectangular shapes of from 5/8" to 1 1/4" wall thicknesses, and in 24" to 30" lengths (Figure 82). The terra

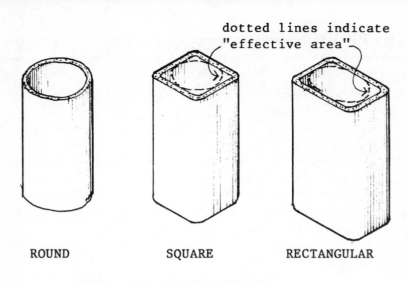

TERRA-COTTA FLUE LININGS
FIGURE 82

cotta lining is constructed within, and supported by, the masonry envelope (Figure 83).

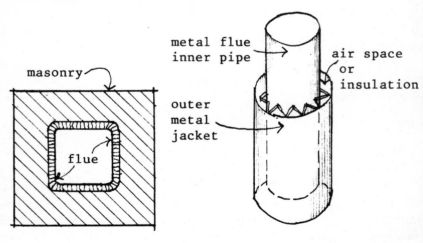

MASONRY CHIMNEY WITH
TERRA-COTTA FLUE LINING

DOUBLE WALL
METAL CHIMNEY

FIGURE 83

(**CAUTION**) A masonry chimney without terra cotta flue linings should not be accepted. Rough inner masonry surfaces without linings would restrict and inhibit the flow of gasses; and, there is also a much greater danger of hot gasses escaping outward thru the masonry.

Metal chimneys are round, made of steel or stainless steel, with an inner wall and an outer wall separated either by an air space or an insulating filler. The separation is necessary to reduce the temperature of the outer wall to acceptable levels. They come in sections of various diameters and lengths. See (Figure 83)

The required cross-sectional area of a flue is dictated by the draft requirements of the heating appliance or fireplace connected to it. A general rule of thumb for a fireplace is that the effective area of the flue should be from a minimum of 1/12th to 1/10th of the face area of the fireplace face opening. The effective area of a round flue is the same as its actual inside area. However, for square or rectangular flues, the effective area is less than the actual area because rising gasses confine themselves to **circular** patterns, and therefore the corners of square or rectangular flues are ineffective. See (Figure 82).

(**CAUTION**) Every heating appliance or fireplace should have **its own separate flue**. One flue should not be used for more than one appliance.

Flues should generally be run vertically straight up to the outlet. Any changes in the direction of the flue must be less than 45 degrees—preferable not over 30 degrees—from the vertical. Sharp bends or turns seriously impact the flow of gasses (Figure 84).

Surrounding a flue, masonry which is exposed to the exterior weather should not be less than 8 inches thick. For interior chimneys this may be reduced to 4 inches if of solid brick. All wood framing and finishes must be separated from masonry chimneys by a minimum of 2 inches, with the space filled with non-combustible insulation. The clearances required to wood for pre-fab metal chimneys may be greater than 2 inches. Consult the data published by the manufacturer of the metal chimney, as well as your local Code.

A masonry chimney has considerable weight and must have a separate footing; most pre-fab metal chimneys or flues do not. Metal chimneys rest on the fireplace unit or appliance which they

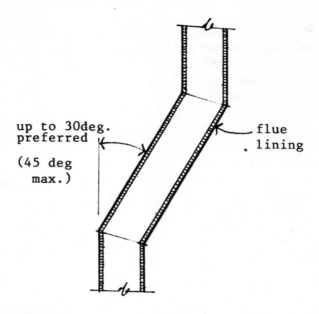

ELEVATION OF CHANGE-IN-DIRECTION OF FLUE

FIGURE 84

serve, but must be stabilized or braced at frequent intervals so that they do not collapse or dis-locate sideways. Wherever a chimney or flue penetrates a floor, ceiling or roof, the space between the chimney and the floor, ceiling or roof must be firestopped, that is, closed up with non-combustible materials. Special metal *firestop spacers* are made for this purpose by the manufacturers of most pre-fab chimneys.

A chimney must be flashed to the roof at its penetration thru it, or wherever it contacts the roof, if it is on an outside wall. See (Figures 57 and 81).

FIREPLACES:

A. MASONRY FIREPLACE is one built entirely of masonry materials at the jobsite, without the use of pre-fabricated metal forms or liners. Since they are custom or job-built, masonry fireplaces therefore can be built in a wide variety of sizes, shapes and types, in contrast to those fireplaces constructed around metal forms or liners which are available only in limited sizes and types.

There are several critical factors and principles which influence the proportions and details of a masonry fireplace, as follows, See (Figure 85):

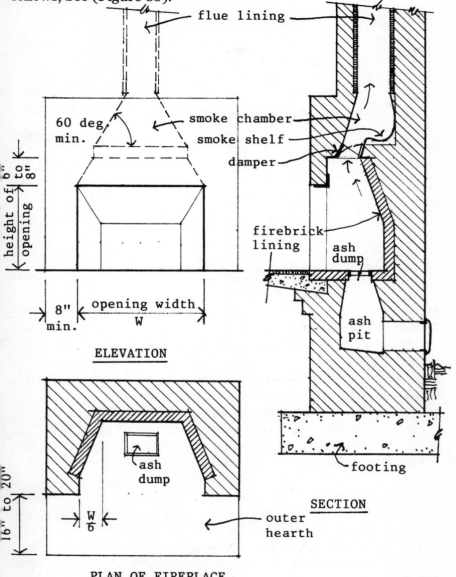

BASIC DETAILS OF MASONRY FIREPLACES

FIGURE 85

1) The size and shape of the fireplace opening. Openings should be rectangular, wider than high, the height being 2/3 to 3/4 of the width. Some of the more common sizes are: 32"w x 22" to 24"h; 36"w x 24"h; 42"w x 26" to 30"h; 48"w x 28" to 32"h.
2) The flue size must be proportioned based on the size of the fireplace opening. The effective area of the flue being a min of 1/12th to 1/10th of the fireplace face opening area.
3) The front-to-back depth of the fireplace should not be too great, generally within 1/2 to 2/3rd of the height. 18" to 20" works for fireplaces up to 42" wide; 21" to 22" for width of 48".
4) The sides of the fireplace should taper or slope inward from front to back, at approx. 5 inches per foot of depth, and/or 1/6th of the opening width, each side.
5) The fireplace back wall should taper or slope inward towards the top, or *throat*, of the fireplace.
6) The throat should be located slightly above (6 to 8 inches) the top of the front opening, and should contain a metal damper, which is the full width of the throat. The damper is to control the *draft* or flue action of the fireplace, as well as to prevent cold air downdrafts when the fireplace is not in use.
7) A *smoke chamber* should be created just above the damper, which has its sides tapered at a 60 degree angle towards the bottom of the terra cotta flue. The base of the smoke chamber should be curved to deflect cold downdrafts upwards. This is called the *smoke shelf*. The warm air rises in a column into the flue, while at the same time there is a downward current of cool air close to the flue walls. Without the deflecting upward action of the smoke shelf, this cool air downdraft would block the smoke and gasses at the throat, forcing them back into the room.
8) There should be a hearth of non-combustible material at the fireplace floor level, which projects out beyond the fireplace opening approx. 16 to 20 inches, and several inches (usually 8") wider than the width of the fireplace opening. This is a precaution against flying sparks and embers which would otherwise set combustible floors on fire.
9) All exposed inner surfaces of a masonry fireplace should be constructed of special "fire brick".

Oftentimes, the area beneath the floor, or back hearth, of the fireplace will be created into a hollow chamber as an ash storage area. An outside *cleanout* door is built in as access to the ash storage area. A metal *ash dump* device, or door, will be built into the fireplace floor for this purpose.

See (Figure 85) for graphic illustrations of these components and principles.

Masonry fireplaces can be built with openings on the front and one side, or openings front and back. Such fireplaces have special design rules and concerns which are considered to be beyond the purposes of this book. The reader is urged to conduct some research into the details of special fireplaces if any are being considered.

B. PRE-FABRICATED FIREPLACES: Metal forms or liners in certain stock sizes are available from various manufacturers (Heatilator, Temco, Heat-N-Glo, etc.) which take the place of the job-constructed inner surfaces of the fireplace. These can be set in place on a non-combustible base. Masonry or other materials are used to enclose them, to give the appearance of a normal fireplace. There are types which must have masonry materials surrounding and enclosing them; and, there are types using multi-wall, insulated construction which do not require masonry, and can be installed with very minimal clearances to combustible materials. In all cases, the Code and the manufacturers written instructions must be followed. See (Figure 86).

C. FREE-STANDING FIREPLACES AND STOVES: Various companies manufacture metal-free standing fireplaces and wood burning stoves, in a variety of designs and profiles. These must rest on non-combustible materials, and all have certain minimum clearance requirements to combustible materials as well as furnishings. Codes and manufacturers written specifications must be followed in their installation. See (Figure 87).

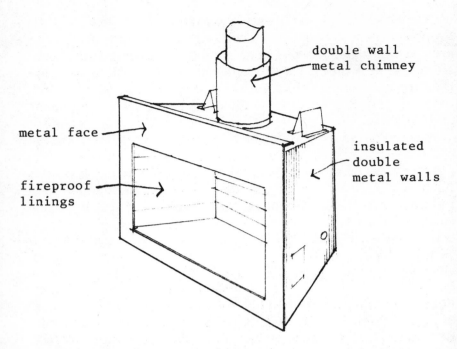

LOW CLEARANCE, BUILT-IN TYPE METAL FIRE PLACE

FIGURE 86

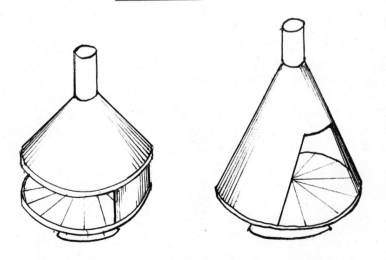

EXAMPLES OF METAL FREE-STANDING FIREPLACES

FIGURE 87

APPENDIX 'A'

STATE LISTING OF MODEL BUILDING CODES IN USE

Key No.	Applicable Code, and Extent of Use
1	NATIONAL BUILDING CODE (portions only) by Building Officials and Code Administrators International, Inc. Statewide usage.
2	NATIONAL BUILDING CODE, Some Local Jurisdictions only.
3	NATIONAL BUILDING CODE, State Buildings only.
4	NATIONAL BUILDING CODE, Manufactured Housing only.
5	UNIFORM BUILDING CODE by International Conference of Building Officials, Adopted as State Building Code or is used on a statewide basis.
6	UNIFORM BUILDING CODE, State Buildings Only.
7	UNIFORM BUILDING CODE, Housing only.
8	UNIFORM BUILDING CODE, Some Local Jurisdictions Only.
9	STANDARD BUILDING CODE, by Southern Building Code Congress International, Inc.
10	Has its Own Building Code, available for local jurisdiction adoption or use.

STATES:

ALABAMA: 9
ALASKA: 5
ARIZONA: 5
ARKANSAS: 8 & 9
CALIFORNIA: 7
COLORADO: 5
CONNECTICUT: 1 & 2
DELAWARE: 1, 2 & 9
FLORIDA: 9
GEORGIA: 9
HAWAII: 5
IDAHO: 5
ILLINOIS: 1, 2 & 8
INDIANA: 2 & 5
IOWA: 5
KANSAS: 2, 8 & 9
KENTUCKY: 1 & 2
LOUISIANA: 8 & 9
MAINE: 2 & 3
MARYLAND: 2, 4 & 9
MASSACHUSETTS: 1 & 2
MICHIGAN: 1, 2 & 8
MINNESOTA: 5
MISSISSIPPI: 9
MISSOURI: 2 & 8
MONTANA: 5
NEBRASKA: 2 & 6
NEVADA: 5
NEW HAMPSHIRE: 2 & 3
NEW JERSEY: 1 & 2
NEW MEXICO: 5
NEW YORK: 2 & 10
NORTH CAROLINA: 9
NORTH DAKOTA: 2 & 8
OHIO: 1 & 2
OKLAHOMA: 1, 2, 8 & 9
OREGON: 5
PENNSYLVANIA: 2 & 4
RHODE ISLAND: 1 & 2
SOUTH CAROLINA: 9
SOUTH DAKOTA: 8
TENNESSEE: 9
TEXAS: 2, 8 & 9
UTAH: 5
VERMONT: 1 & 2
VIRGINIA: 1 & 2
WASHINGTON: 5
WEST VIRGINIA: 2, 8 & 9
WISCONSIN: 1 & 2
WYOMING: 5

MINIMUM SEPTIC TANK CAPACITIES FOR SINGLE FAMILY DWELLINGS

BEDROOMS SERVED[1]	MINIMUM TANK LIQUID CAPACITY (Gallons)
1-3	960
4	1,200
5	1,500
6[2]	1,800

[1]Dens and garages that can be converted to bedrooms may be included at the discretion of the county health department
[2]For more than six bedrooms, use 1.6 x 200 x number of bedrooms for minimum tank capacity in gallons

MINIMUM SETBACK REQUIREMENTS FOR SEPTIC-TANK SYSTEMS[4]

	SEPTIC TANK	DISPOSAL TRENCH	DISPOSAL PIT
Buildings	10 feet	10 feet[3]	10 feet[3]
Property lines[1]	5 feet	5 feet	5 feet
Wells (Public Water Supplies)	100 feet	100 feet	100 feet
Wells (Private)	50 feet	50 feet	50 feet
Live streams[2]	100 feet	100 feet	100 feet
Dry wash	50 feet	50 feet	50 feet
Water lines	10 feet	10 feet	10 feet
Cuts on sloping terrain		50 feet	50 feet

[1]Lots with individual wells require setbacks of 50 feet
[2]200 feet on water supply watersheds
[3]Or minimum spacing between trenches (see Part V, B, 9) or disposal pits (see Part VI, B, 12) whichever is greater
[4]All distances are from edge to edge

APPENDIX 'B'

SEPTIC SYSTEM DATA FOR PIMA COUNTY, ARIZ.
(Reprinted with permission of State of Arizona, Department of Environmental Quality)

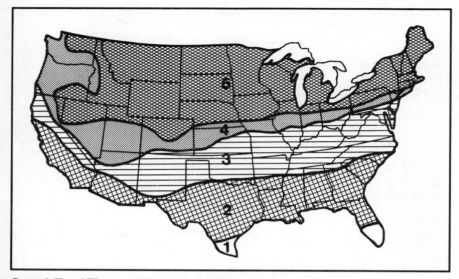

CertainTeed Thermal Recommendations

Homes not insulated to today's energy standards can mean substantial heat loss in winter and heat gain in summer. Installing proper amounts of thermally efficient fiber glass insulation is one of the most cost-effective energy conservation measures that can be taken.

Climate and fuel costs determine how much insulation should be used for maximum economic return. As fuel and electric costs increase, higher R-Values are usually justified.

In some areas of the country, heating costs determine how much insulation should be used. In other areas air-conditioning costs are the major influence. In many areas *both* heating and air-conditioning costs contribute to the insulation recommendations shown here.

Find the appropriate zone. Then use the R-Values on the chart.

Zone	Heating Degree Days	Ceiling R-Value	Wall R-Value	Floor R-Value
1	up to 500	R-19	R-12	R-11
2	501-3,000	R-30	R-19	R-11
3	3,001-5,000	R-38	R-19	R-14
4	5,001-6,000	R-38	R-19	R-22
5	6,001 up	R-49	R-22	R-25

R-Values on the chart represent CertainTeed recommendations based on the new CABO Model Energy Code effective January 1, 1988. It takes into consideration both heating and cooling. The Wall R-Values include those for both insulation and sheathing. If the home has a heated basement, the basement walls should be insulated to the level so that the calculated energy consumed would be no greater than that if the basement was unheated and the floor above insulated to the Floor R-Value indicated.

APPENDIX 'C'

MINIMUM RECOMMENDED 'R' VALUES FOR INSULATION
(Reprinted with permission of the CertainTeed Corp)

GLOSSARY

A.B.C.
Aggregate Base Course, a blended mixture of natural granular materials, such as crushed stone, gravel, slag or crushed ledge rock, which is suitable for use as the non-moisture-retaining base for floor slabs and other pavements.

BEARING
1) To rest on, such as a beam *bearing* on a wall; 2) to support floor or roof structural loads in addition to its own weight, such as a *bearing* wall.

NON-BEARING
A wall or partition which does not support loads such as floor or roof members, other than its own weight.

BACKFILL
The process of replacing soil into an excavated cavity, such as a trench or a foundation.

BOND
The physical connection of one material to another, usually relating to achievement of bond during embedment, such as the bond of steel re-bars to concrete in which they are embedded. Also, the bonding of masonry units by staggering vertical joints in alternate courses.

B.T.U.
BRITISH THERMAL UNIT. A measurement of the amount of heat required to raise the temperature of one pound of pure water one degree Farenheit.

COLUMN
A free-standing, vertical, structural support member. May be of wood, concrete, masonry, or steel. Commonly seen are round steel pipe, wood 4"x4", brick or concrete block piers, etc.

COMPACTION
The process of applying weight pressure and vibration to soil which has been disturbed, excavated or moved, to restore the soil to an acceptable range of its original un-disturbed density.

COURSE
A single horizontal layer of masonry units, such as a *course* of brick, a *course* of block, etc.

CURING
The process of protecting newly placed concrete from the effects of too low or too high temperatures, and premature dry-out.

DAMPPROOFING
Protection against the effects of moderate wetness or dampness. Less extensive in make-up and effectiveness than Waterproofing.

EFFLUENT
Water containing contaminates, which is discharged into, or from, sewage systems.

FASCIA
The face trim member which covers the outer ends of roof rafters or joists.

FILL
Soil excavated or moved from one location and redeposited at another location by man, for purposes of: 1) Changing topography or grades; 2) Improving or modifying soil drainage characteristics; or 3) modifying or improving soil load-bearing capacities.

FIRESTOPPING
The closing up of open vertical chases or shafts to prevent fire and smoke from penetrating from one floor, level or compartment into the next.

FLASHING
An assembly of waterproof materials which seals and makes watertight the intersections of dissimilar materials, roof planes, roofs and intersecting walls, and penetrations thru roofing.

FLUE
The inner open chamber of a chimney, which contains and conducts the actual hot gasses.

FOOTING
The below-grade structural components, which receive all the loads from the upper portions of the building and transmit those loads safely and uniformly to the soil.

FOUNDATION
The combination of below-grade walls and footings which support the above-grade portions of the building.

FOUNDATION WALL
The below grade walls which transmit the loads of the superstructure above, to the footings.

FRAME
A construction system consisting of a series of repetitive members forming a structural skeleton of the walls, floors and roof of a building.

GRADE
The ground surface, either in its natural or its altered state. Also refers to the process of establishing ground surfaces, as in *grading* the site to improve drainage.

GROUT
A cement-sand-water mix which is poured into voids or cavities in masonry to add strength and to surround reinforcing steel, thus bonding it to the masonry.

JOISTS
The repetitive structural support members in floor and roof framing.

LATERAL
Sideways; as opposed to up-and-down, or vertical.

LOADS
The forces placed on members or surfaces due to: 1) The dead weight of the materials 2) The weight of occupants and furnishings (called *live* loads); and 3) wind and/or earthquake action, also considered as *live* loads.

LOAD BEARING CAPACITY
The degree of capability a material possesses to support loads placed upon it, without failure or collapse.

MORTAR
A masonry cement-sand-water mix which is placed in all horizontal and vertical joints to cement, or bond, together adjacent masonry units, and to make the joints resistant to the passage of water and wind.

PARTITION
A non-load bearing wall, for the purposes of dividing, sub-dividing or enclosing space.

PERCOLATION
The rate at which water can naturally penetrate into, or thru, soil.

PITCH
A sloping away from dead-level horizontal, to cause run-off of water.

PLATE
The base, and the top cap, members of stud partitions to which the studs are connected thereby aligning and positioning the studs. Also, refers to the wood *sill* members which are anchored to foundation walls as a base for floor or walls above to rest upon and attach to.

PLY
In roofing, an individual layer of the overall roofing materials system. Example: Built-up asphalt roofs may have from 3 to 5 **plies** of special roofing felts with hot asphalt between each ply.

PSI
PER SQUARE INCH. Usually used in describing the tensile, bending and compressive strength of a material. Expressed in number of pounds per square inch (lbs./psi).

RE-BAR
Deformed steel reinforcement bar. Solid steel members, usually in round bar configurations, made in various diameters and long lengths, which have special ridges and patterns projecting from their surface for purposes of achieving bond to the concrete in which the re-bar is embedded.

SHEATHING
The material applied to the outer faces of stud walls and joist roofs, as exterior enclosure and as a base for the attachment and support of the finish exterior materials.

SILL
The continuous member(s) anchored to the top of foundation walls as a base for floor or wall systems above to rest upon and attach to.

SITE
The portion of land dedicated to the particular building activity.

SOFFIT
The underside of roof overhangs.

SOLE
The base member, or *plate*, to which upper members are aligned and rest upon. Typically refers to the 2"x base sole/plate of wood stud partitions.

SQUARE
In roofing refers to 100 square feet of surface. In geometry and layout, refers to angles being exactly 90 degrees.

STEM WALL
The foundation wall between a grade-supported floor slab and the footing.

STUCCO
A plaster-like exterior finish material made of portland cement, lime, sand and water.

STUD
The vertical members, usually extending from floor to ceiling, which are the structural support elements within a wall or partition. They may be of wood or of metal, and are placed at some set spacing throughout the length of the wall. The surface finish materials are attached to the studs.

SUB-FLOORING
The first layer structural floor decking, to which is applied the finish flooring material.

SUB-SURFACE
Refers to conditions found or created below the surface of the ground.

SUMP SYSTEM
A combination of a basin or container for water to flow to and collect in, plus a means such as a mechanical pump, for discarding the water to an outside drain or drainage system. A relatively common system used for disposing of water in excavated basements.

SUPER-IMPOSED LOADS
The loads applied to a surface or a member due to the materials and occupant loads above.

SWALE
A ditch, either naturally occuring or man made, for the purposes of intercepting and diverting surface water drainage.

TOPOGRAPHY
The surface configuration, elevations and pitch of the ground.

U.L.
UNDERWRITERS LABORATORIES, an independent testing laboratory which determines the degree of fire hazard or fire resistance qualities of materials and assemblies of materials.

UNDERLAYMENT
An additional layer of material which provides the proper sub-surface quality and density required for the installation of certain finish flooring materials. The basic sub-flooring may, or may not, be of suitable *underlayment* quality. Particle board is often used for underlayments. Certain plywood can be obtained in *underlayment grade.*

VAPOR BARRIER
A continuous, un-interrupted sheet or membrane which has the ability to resist the passage thru it of water in both liquid and vapor (gas) form.

VENEER
A non-structural surfacing or *skin* used for its particular appearance. Usually refers to masonry (brick, stone, etc.) veneer on the exterior walls of a building.

WATERPROOFING
A process or assembly of materials intended to prevent the penetration of standing water, and withstand frequent exposure to moisture.

WATER TABLE
The level, or elevation, in the ground below which the soil is saturated with standing water.

INDEX

A.B.C. 42
Adobe 7
Air Infiltration 77
Anchorage 37,50,61
Anchors, framing 50
Asphalt shingles 101

Backfill 16
Basement 7
Basement Floors 41
Basic Criteria affecting
   selection of roofing 95
Bearing 56,61,115
Bracing 65
Bridging................52
Building Codes, Officials....13
Built-up Roofing 103

Cabinets...............139
Cabinetwork and
   Built-ins 139
Carpet 135
Ceramic tile 136
Chairs, re-bar support 30
Chimneys 141
Chimneys and
   Fireplaces.............141
Clay 18
   Tile 105
Compaction of Soil 17
Concrete 32
   Basement 41
   Compressive strength 34
   Curing 34,44
   Floors at grade 42
   Set 34
Control Joint 44
Counter Tops 140
Crawl Space............50

Dampproofing
   and Waterproofing 37
Doors 77
   Batten 78
   Construction and types ..77

Frame 83
Flush 78
Horizontal Sliding 80
Panel 77
Pocket 80
Sectional Overhead......82
Sizes.................78
Sliding Exterior 80
Swinging hinged 80
Drainage 15
Drain, foundation.........40
Drywall Finishes.........124
   Gypsum Board 124
   Plywood, Other 128
   Joint Treatments.......128

Effluent 19
Exterior Walls 61
Expansion Joint..........44

Fill....................17
Fireplaces 144
   Free-Standing 147
   Masonry 144
   Pre-fabricated 147
Firestopping 65
Fire Resistivity..(roofing)..97
Flashing 52,69,95,106
   Basic Principals 106
   Floor 66
   Ice Dam 109
   Parapets..............87
   Sill 52
Floor Systems 41
Flooring 131
   Carpet 135
   Ceramic Tile 136
   Parquet..............134
   Resilient 134
   Strip 133
   Wood 131
Flue, Flue Lining 141
Footings................25
   Continuous Wall 27
   Individual 27

INDEX

Stepped 27
  and Foundations 25
  Forms 31
Forms, footing 31
Foundations 7,35
Framing Anchors and Ties . . 50
Framing Lumber 44
Free-Standing Fireplaces
  and Stoves 147
Frost, freezing 7,26
Furr, Furring 66

Glass 75
  Insulating 75
  Shading Coefficient 76
Grading 16
Gound Water 7,40
Gutters and Leaders 109
Gypsum Board Drywall 124
  Casing bead 127
  Green Board 124
  Joint Treatment 125
  Nailing 124

Hardner 44
Hardware 83
  Latchset 83
  Lockset 83
Hinges, butt 83
  Pin 83

Ice Dam Flashing 109
Insulation 61,69
  Resistance 'R' 69
  'U' value 69,70
Interior Stairs 117
Introduction 3

Joists 45

Keyway 29
Kiln 44

Lath and Plaster Finish 123
  Keenes Cement 123
  Lath 122
  Plaster 123
  Thin Coat 124
Load Bearing Capacity 25
Longevity (roofing) 98
Lumber, drying 44
  Grading 45
  Green 45
  Framing 44

Masonry 65
  Course 35
  Furring 66
  Joint tooling 36
  Thermal Characteristics . . 66
  Unit masonry 35
  Veneer 68
  Walls 65
Membrane
  Waterproofing 39,41
Methods of Operation
  of Doors 80
Millwork 130
Mortised 83
Non-Bearing 56,61,115

Painting 136
Parapets 87
Parquet Flooring 134
Partitions 115
  Rough Openings 115
  Thin 115
Percolation test 19
Performance Code 13
Plate 37,47,61,65
Plaster 123
Platform Frame 45
Preface 1
Pre-Fabricated Fireplaces . . 147
Protection of Plants 18
Plywood, Wood Pulp and
  other Drywall panels 128

INDEX

Re-bar 29,31
Regional Differences 7
Reinforcement 29
Resilient Flooring 134
Resistance, resistivity 69
Retain and Protect Plants . . . 18
Role of Building Codes,
    Officials 13
Roofing 95
    Appearance 95
    Built-up 95,103
    Clay tile 105
    Fire Resistivity 97
    Longevity 98
    Roll 99
    Shingles 97,98,101
    Slate 105
    Warranties 104
Roof Structure 87
    Deck 99
    Overhang 90
    Pitch, or Slope 95
    Shapes of 87
    Wind Uplift 112

Septic systems 19
Sheathing 61
Sill 37,47
Site Preparation 15
Slump, test for 32
Softwood 44
Sole 61
Sound Transmission 128
Specification Code 13
Stairs, Interior 117
    Carriages 117
    Guidelines, proportions . . . 118
    Stringers 117
Stepped Footings 32
Stem Wall 49
Stucco 11,65
Sub-Flooring 50

Termites,
    Subterranean 20,65
    Shields 52
    Treatment for 20
Thermal conduction 69
Topsoil 18
Trim, wood 130
    Moldings 131
Trusses 90
Types of Roofing, and
Installation requirements . . . 99

Underlayment 50,135
'U' value 69,70,76

Vapor Barrier 41,43,50,61
Veneer, masonry 52,68
Ventilation 90
Vertical Re-Bar 31

Wainscots 131
Wall Ties 69
Waterproofing 7,37
    Membrane 39,41
Weatherstripping 77
Weep Holes 52
Wind Forces and Uplift
    on Roofs 112
Windows 73
    Metal 73,75
    Wood 75
Wood Floor 50
    Framing 45
    Flooring 131
Wood Frame 61
Wood Frame with Masonry
Veneer 68
Wood Trim and Millwork . . . 130